INSTRUMENTATION FOR CIVIL ENGINEERING APPLICATIONS

AUTHOR (S)

Prof. Ashim Kanti Dey
Professor
Department of Civil Engineering,
National Institute of Technology
Silchar, Silchar

Dr. Manita Das
Assistant Professor
Department of Civil Engineering,
ITER, SOA (Deemed to be
University), Bhubaneswar

REVIEWER

Prof. Neelima Satyam
Professor
Department of Civil Engineering,
Indian Institute of Technology Indore, Indore

KHANNA BOOK PUBLISHING CO. (P) Ltd.
4C/4344, Ansari Road, Darya Ganj,
New Delhi, India

All India Council for Technical Education
Nelson Mandela Marg, Vasant Kunj,
New Delhi, India

Dear Readers,

To prevent the piracy, this book is secured with HIGH SECURITY HOLOGRAM on the front title cover. In case you don't find the hologram on the front cover title, please write us to at contact@khannabooks.com or whatsapp us at +91-99109 09320 and avail special gift voucher for yourself.

Specimen of Hologram on front Cover title:

Moreover, there is a SPECIAL DISCOUNT COUPON for you with EVERY HOLOGRAM.

How to avail this SPECIAL DISCOUNT:

Step 1: Scratch the hologram

Step 2: Under the scratch area, your "coupon code" is available

Step 3: Logon to www.khannabooks.com

Step 4: Use your "coupon code" in the shopping cart and get your copy at a special discount

Step 5: Enjoy your reading!

ISBN: 978-93-5538-323-5

Book Code: UG434EN

Instrumentation For Civil Engineering Applications

by Prof. Ashim Kanti Dey and Dr. Manita Das

First Edition: 2026

Originally Published by:

All India Council for Technical Education, Ministry of Education, GOI

Exclusively Published & Distributed by:

Khanna Book Publishing Co. (P) Ltd.

4C/4344, Ansari Road, Daryaganj,
New Delhi (INDIA)

Visit us at: **www.khannabooks.com**

Write us at: **contact@khannabooks.com**

To view complete list of books, Please scan the QR Code:

KPH

Printed in Bharat.

प्रो. टी. जी. सीताराम
अध्यक्ष
Prof. T. G. Sitharam
Chairman

सत्यमेव जयते

आज़ादी का अमृत महोत्सव

अखिल भारतीय तकनीकी शिक्षा परिषद्
(भारत सरकार का एक सांविधिक निकाय)
(शिक्षा मंत्रालय, भारत सरकार)
नेल्सन मंडेला मार्ग, वसंत कुंज, नई दिल्ली–110070
दूरभाष : 011–26131498
ई–मेल : chairman@aicte-india.org

ALL INDIA COUNCIL FOR TECHNICAL EDUCATION
(A STATUTORY BODY OF THE GOVT. OF INDIA)
(Ministry of Education, Govt. of India)
Nelson Mandela Marg, Vasant Kunj, New Delhi-110070
Phone : 011-26131498
E-mail : chairman@aicte-india.org

FOREWORD

Engineers are the backbone of any modern society. They are the ones responsible for the marvels as well as the improved quality of life across the world. Engineers have driven humanity towards greater heights in a more evolved and unprecedented manner.

The All India Council for Technical Education (AICTE), have spared no efforts towards the strengthening of the technical education in the country. AICTE is always committed towards promoting quality Technical Education to make India a modern developed nation emphasizing on the overall welfare of mankind.

An array of initiatives has been taken by AICTE in last decade which have been accelerated now by the National Education Policy (NEP) 2020. The implementation of NEP under the visionary leadership of Hon'ble Prime Minister of India envisages the provision for education in regional languages to all, thereby ensuring that every graduate becomes competent enough and is in a position to contribute towards the national growth and development through innovation & entrepreneurship.

One of the spheres where AICTE had been relentlessly working since past couple of years is providing high quality original technical contents at Under Graduate & Diploma level prepared and translated by eminent educators in various Indian languages to its aspirants. For students pursuing 2nd year of their Engineering education, AICTE has identified 87 books, which shall be translated into 12 Indian languages - Hindi, Tamil, Gujarati, Odia, Bengali, Kannada, Urdu, Punjabi, Telugu, Marathi, Assamese & Malayalam. In addition to the English medium, books in different Indian Languages are going to support the students to understand the concepts in their respective mother tongue.

On behalf of AICTE, I express sincere gratitude to all distinguished authors, reviewers and translators from the renowned institutions of high repute for their admirable contribution in a record span of time.

AICTE is confident that these outcomes based original contents shall help aspirants to master the subject with comprehension and greater ease.

(Prof. T. G. Sitharam)

ACKNOWLEDGEMENT

We are deeply grateful to the authorities of AICTE, particularly Prof. T. G. Sitharam, Chairman; Dr. Abhay Jere, Vice-Chairman; Prof. Rajive Kumar, Member-Secretary, and Dr. Sunil Luthra; Director, Training and Learning Bureau, for allowing us to write this book. Their meticulous planning and unwavering support in publishing the book on '**Instrumentation for Civil Engineering Applications**' have been instrumental. We personally thank Mr. Sachin Kumar who has helped us in designing the front pages of the book as per AICTE norms. We also extend my sincere appreciation to the book reviewer, Prof. Neelima Satyam, Professor, Department of Civil Engineering, Indian Institute of Technology Indore for her invaluable inputs across all book units.

This book is a testament to the collective wisdom and shared vision of the Government of India, AICTE members, experts, and authors who have generously shared their insights to advance engineering education in our country. We extend our heartfelt acknowledgements to the numerous contributors in this field whose published books, review articles, papers, photographs, footnotes, references, advice, and other valuable information have enriched our work.

We wish to acknowledge the support and great love of our family members who gave us continuous love and support to write this book.

Ashim Kanti Dey
Manita Das

PREFACE

It is our great pleasure to write a book on a topic which is not much explored by the Civil Engineers but is used in all precise field and research works. All the sensors or instruments that we use for our work are simple in construction and easy in operation. The theory behind the sensors consists of simple electrical circuits which all of us have studied in class XI – XII standards.

This book aims to provide a holistic overview of instrumentation for civil engineering applications. At the beginning we shall study the different instruments which are being used for measuring the basic physical quantities like force, displacement, pressure, inclination, etc. The measuring instruments are constructed with simple electrical theories like generate an electric potential, measure the current and obtain the resistance. However, measuring the signal is an art, since a precise measurement will give an accurate result. In actual field, the construction of a sensor is not so simple but is also not complicated. Throughout the book, we have tried to provide an easy demonstration to the manufacturing process of the instruments, collection of data, reduction and interpretation of data.

The contents of the book are normally not taught in most of the Technical Institutes in India although all the instruments are regularly being used by the Civil Engineers. Here lies the importance of the book. We think AICTE has meticulously designed the syllabus and appropriately introduced the subject to the undergraduate level. We feel proud to get this opportunity to give a shape to our unexplored study area.

Lastly, we do hope that the readers of this book will be inspired to fabricate the instruments themselves and will explore the fundamentals of collection of data, interpretation of data and practical applications of the interpreted data. This will revolutionise the design concepts of structure, performance of structure and finally will reduce the high factor of safety.

Ashim Kanti Dey

Manita Das

OUTCOME BASED EDUCATION

For the implementation of an outcome based education the first requirement is to develop an outcome based curriculum and incorporate an outcome based assessment in the education system. By going through outcome based assessments evaluators will be able to evaluate whether the students have achieved the outlined standard, specific and measurable outcomes. With the proper incorporation of outcome based education, there will be a definite commitment to achieve a minimum standard for all learners without giving up at any level. At the end of the programme running with the aid of outcome based education, a student will be able to arrive at the following outcomes:

PO1. Engineering knowledge: Apply the knowledge of mathematics, science, engineering fundamentals, and an engineering specialization to the solution of complex engineering problems.

PO2. Problem analysis: Identify, formulate, review research literature, and analyse complex engineering problems reaching substantiated conclusions using first principles of mathematics, natural sciences, and engineering sciences.

PO3. Design / development of solutions: Design solutions for complex engineering problems and design system components or processes that meet the specified needs with appropriate consideration for the public health and safety, and the cultural, societal, and environmental considerations.

PO4. Conduct investigations of complex problems: Use research-based knowledge and research methods including design of experiments, analysis and interpretation of data, and synthesis of the information to provide valid conclusions.

PO5. Modern tool usage: Create, select, and apply appropriate techniques, resources, and modern engineering and IT tools including prediction and modeling to complex engineering activities with an understanding of the limitations.

PO6. The engineer and society: Apply reasoning informed by the contextual knowledge to assess societal, health, safety, legal and cultural issues and the consequent responsibilities relevant to the professional engineering practice.

PO7. Environment and sustainability: Understand the impact of the professional engineering solutions in societal and environmental contexts, and demonstrate the knowledge of, and need for sustainable development.

PO8. Ethics: Apply ethical principles and commit to professional ethics and responsibilities and norms of the engineering practice.

PO9. Individual and team work: Function effectively as an individual, and as a member or leader in diverse teams, and in multidisciplinary settings.

PO10. Communication: Communicate effectively on complex engineering activities with the engineering community and with society at large, such as, being able to comprehend and write effective reports and design documentation, make effective presentations, and give and receive clear instructions.

PO11. Project management and finance: Demonstrate knowledge and understanding of the engineering and management principles and apply these to one's own work, as a member and leader in a team, to manage projects and in multidisciplinary environments.

PO12. Life-long learning: Recognize the need for, and have the preparation and ability to engage in independent and life-long learning in the broadest context of technological change.

COURSE OUTCOMES

After completion of the course the students will be able to:

CO-1: To analyze the errors during measurements

CO-2: To specify the requirements in the calibration of sensors and instruments

CO-3: To describe the noise added during measurements and transmission

CO-4: To describe the measurement of electrical variables

CO-5: To describe the requirements during the transmission of measured signals

CO-6: To construct Instrumentation/Computer Networks

CO-7: To suggest proper sensor technologies for specific applications

CO-8: To design and set up measurement systems and do the studies

Course Outcomes	Expected Mapping with Programme Outcomes *(1- Weak Correlation; 2- Medium correlation; 3- Strong Correlation)*											
	PO-1	**PO-2**	**PO-3**	**PO-4**	**PO-5**	**PO-6**	**PO-7**	**PO-8**	**PO-9**	**PO-10**	**PO-11**	**PO-12**
CO-1	3	1	2	1	1	1	-	-	2	1	-	2
CO-2	3	3	3	2	-	-	-	1	2	-	1	1
CO-3	3	3	3	2	2	-	-	-	1	1	-	1
CO-4	3	2	3	1	3	1	-	2	2	1	1	3
CO-5	3	2	2	2	3	-	1	1	1	2	-	1
CO-6	3	1	2	1	3	-	1	-	3	-	2	3
CO-7	3	1	2	3	1	1	-	1	1	1	2	-
CO-8	3	2	3	2	1	1	1	1	1	1	1	2

GUIDELINES FOR TEACHERS

To implement Outcome Based Education (OBE) knowledge level and skill set of the students should be enhanced. Teachers should take a major responsibility for the proper implementation of OBE. Some of the responsibilities (not limited to) for the teachers in OBE system may be as follows:

- Within reasonable constraint, they should manoeuvre time to the best advantage of all students.
- They should assess the students only upon certain defined criterion without considering any other potential ineligibility to discriminate them.
- They should try to grow the learning abilities of the students to a certain level before they leave the institute.
- They should try to ensure that all the students are equipped with the quality knowledge as well as competence after they finish their education.
- They should always encourage the students to develop their ultimate performance capabilities.
- They should facilitate and encourage group work and team work to consolidate newer approach.
- They should follow Blooms taxonomy in every part of the assessment.

Bloom's Taxonomy

Level	Teacher should Check	Student should be able to	Possible Mode of Assessment
Create	Students ability to create	Design or Create	Mini project
Evaluate	Students ability to justify	Argue or Defend	Assignment
Analyse	Students ability to distinguish	Differentiate or Distinguish	Project/Lab Methodology
Apply	Students ability to use information	Operate or Demonstrate	Technical Presentation/ Demonstration
Understand	Students ability to explain the ideas	Explain or Classify	Presentation/Seminar
Remember	Students ability to recall (or remember)	Define or Recall	Quiz

GUIDELINES FOR STUDENTS

Students should take equal responsibility for implementing the OBE. Some of the responsibilities (not limited to) for the students in OBE system are as follows:

- Students should be well aware of each UO before the start of a unit in each and every course.
- Students should be well aware of each CO before the start of the course.
- Students should be well aware of each PO before the start of the programme.
- Students should think critically and reasonably with proper reflection and action.
- Learning of the students should be connected and integrated with practical and real life consequences.
- Students should be well aware of their competency at every level of OBE.

LIST OF FIGURES

LIST OF TABLES

CONTENTS

1 Fundamental of Measurement, Sensing and Instrumentation

UNIT SPECIFICS

Through this unit we have discussed the following aspects:

- *Basic definitions of measurement, instrumentation and sensors;*
- *Different types of sensors;*
- *Difference between a sensor and a transducer;*
- *Different types of instruments;*
- *Different terminologies used in instrumentation;*
- *Interpretation of signal from a sensor*
- *Installation of sensors*

The practical applications of the topics are discussed for solving further curiosity and creativity as well as improving problem solving capacity.

Besides giving a large number of multiple-choice questions as well as questions of short and long answer types marked in two categories following lower and higher order of Bloom's taxonomy, assignments through numerical problems, a list of references and suggested readings are given in the unit so that one can go through them for practice.

RATIONALE

This fundamental unit on instrumentation and sensors helps students to get a primary idea about the different types of instruments and sensors used in Civil Engineering structures. It explains basic definitions of measurement and instrumentation, physical variables that are to be measured through sensors, difference between an instrument and a sensor, different terminologies used in instrumentation. Various types of instrument used in the Civil Engineering field are also discussed in short. A brief note is also given to understand how a signal from a sensor is interpreted. Installation of different sensors in important Civil Engineering Structures is also shown through pictorial diagrams. An appendix is added at the end of the chapter to help you understand the types of errors used in the measurement.

PRE-REQUISITES

Mathematics: General Mathematics (Class XII)

Physics: Engineering Physics (First year, B.Tech)

UNIT OUTCOMES

The list of outcomes of this unit is as follows:

U1-O1: Describe basic definitions of measurement, instrumentation and sensors.

U1-O2: Describe different types of sensors and instruments

U1-O3: Explain the interpretation of the signal from a sensor

U1-O4: Realize the difference between a sensor and an instrument

U1-O5: Analyse errors during measurements

Unit-1 Outcomes	EXPECTED MAPPING WITH COURSE OUTCOMES (1- Weak Correlation; 2- Medium correlation; 3- Strong Correlation)							
	CO-1	CO-2	CO-3	CO-4	CO-5	CO-6	CO-7	CO-8
U1-O1	-	1	1	1	1	-	3	1
U1-O2	1	1	2	-	1	1	1	-
U1-O3	-	-	-	1	1	1	-	-
U1-O4	-	-	-	1	1	1	-	1
U1-O5	3	1	-	-	-	-	-	1

1.1 DEFINITION OF MEASUREMENT

The term 'measure' is obtained from the Latin word 'mensurare'. The Oxford dictionary explains the term measure as "ascertain the size, amount or degree of (something) by using an instrument or device marked in standard units or by comparing it with an object of known size". However, the definition is not complete. The above definition contains two terms (1) estimation and (2) Standard unit. The third term which is very important in the task of measurement, is the uncertainty of estimation. Thus, following definition holds good:

"Measurement is the estimation of the quantity of certain value (with known uncertainty) by comparison with the standard unit" (Tumanski, 2006). Almost all our activities are related to measurement. Suppose we go to any supermarket and wish to pay for the chosen material. Thus, we have estimated the chosen material after comparing it with the standard materials. The uncertainty in the payment lies on absent mindedness or poor eye sight etc.

The quantities which fall under mechanical measurement includes (a) pressure (b) displacement (c) temperature (d) fluid flow, (e) mass, length, time, etc. In modern technology the above quantities are not measured purely by mechanical means but are measured by electrical and electronic means. The measurement process can be shown through Fig. 1.1

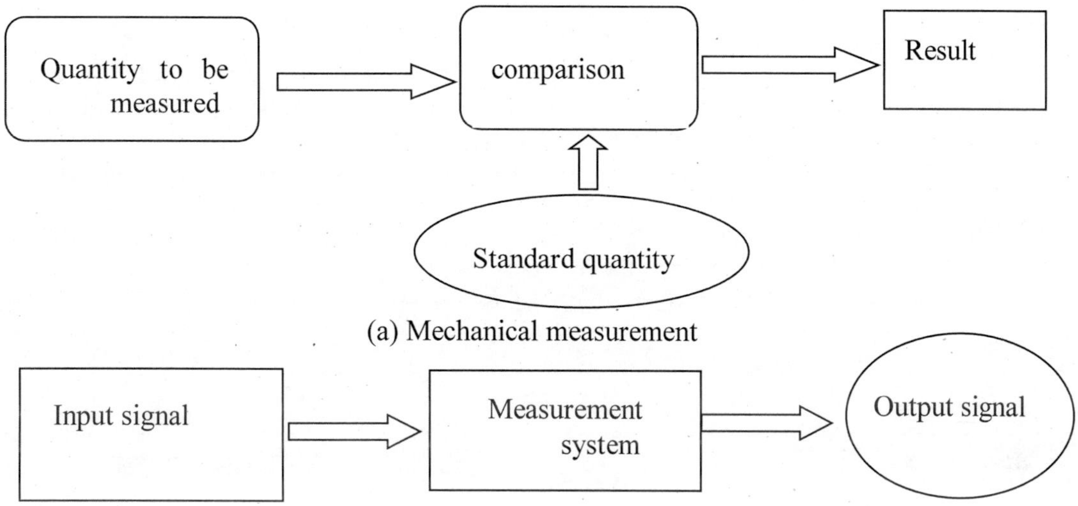

(a) Mechanical measurement

(b) Electrical measurement

Fig 1.1: Measurement process of mechanical and electrical systems

1.1.1 Methods of measurement

The methods of measurements are broadly classified into two categories:

a) Direct method and b) Indirect method

In the direct method the unknown quantity (also called the measurand) is directly compared against a standard quality. Suppose one wants to measure the length of a bar, the bar is to be measured against a standard length, and the outcome will be a measurable quantity, say metre.

The indirect method involves an indirect way of measurement that does not involve human factors. The above length measurement can be done indirectly by some sensor that senses the value of the length and sends the signal to an instrument that shows the measurement digitally or electronically.

1.2 DEFINITION OF INSTRUMENTATION

The instruments are defined as some devices that help in the measurement of the measured quantities or variables. The instruments are thus extensions of human faculties. Some instruments are simple and directly give the measured value whereas some instruments may be complex and consist of a number of separate elements called a system. The system may consist of a preliminary unit which transduces the measurand to an analogous form which is then processed by some intermediate units and is finally transferred to the display unit. The complete system is then called a measurement system.

1.2.1 Development of instruments

The present instruments show three phases of development. The first phase involved mechanical measurements, which were mainly used for static and stable conditions. However, they can not be measured in the dynamic condition. The second phase was electrical instruments which involves mechanical movement. However, the electrical instruments are also time dependent. Most electrical instruments show full scale movement within 0.5 to 24 seconds. Some galvanometers can measure 50 Hz, but modern measurements need fast measurement. Hence the third phase of development is the introduction of electronic instruments. These instruments mainly use semiconductor devices which deploy the movement of electrons. The electrons are excited in a very short time because of their very low inertia. For example, a cathode ray oscilloscope is capable of measuring changes in the electron movement in the order of nanosecond (10^{-9} sec).

1.2.2 Classification of instruments

The instruments are broadly classified into two divisions: absolute and secondary. In absolute instruments the measurement is obtained in terms of the physical constants of the instrument. For example, the probing ring is an absolute instrument where the load is obtained by multiplying the observed readings by the instrument constant. The secondary instruments give direct measurements. The output of the secondary instruments is to be calibrated against some absolute instruments or another secondary instrument. A glass thermometer is an example of a secondary instrument.

1.3 PHYSICAL VARIABLES USED IN INSTRUMENTATIONS

The physical variables which are used in instrumentation consist of

a) Geometric
b) Force-torque
c) Kinematic
d) Thermal
e) Fluid

Table 1.1 shows a brief idea of different physical variables and the type of instrument used to measure the variable:

Table 1.1: Type of instrument based on physical variables

Physical variable	Type of instrument	Physical quantity to be measured
Geometry	Ultrasonic pulse time	Displacement
	Linear variable differential transformer	Displacement
	Strain gauge	Strain

Physical variable	Type of instrument	Physical quantity to be measured
Force-torque	Load cell	Load, torque
Kinematic	Accelerometer Velocity meter	Acceleration velocity
Thermal	Thermal flux sensor Thermometer	Heat flux Temperature
Fluid	Flow meter Pressure transducer	Fluid flow Pressure

1.4 SENSORS

According to the Instrument Society of America the definition of a sensor is "a device which provides a usable output in response to a specified measurand". The measurand is also termed as stimulus which is the quantity, property or condition that is received by the sensor. The output may be an electrical quantity of a mechanical or optical signal. Another definition of a sensor is that it senses a variation of input energy and produces a variation in another or the same form of energy. Thus, piezoelectric crystal may be considered as a sensor which converts mechanical energy to electrical energy and vice versa.

How do sensors work ?

A sensor converts a physical phenomenon into an analog voltage which is measurable and is displayed in the form of readings for further processing.

Selection of sensor

The civil engineers are constantly using sensors to monitor, measure and assess the quality of construction. Since the subject has not been introduced in the standard curriculum hence it is difficult to select a proper sensor for a typical construction. The following points should be kept in mind during the selection process:

- Accuracy of the sensor
- Precision of measurement
- Appropriate cost of the sensor
- Proper installation of the sensor
- Any special requirement

A sensor is basically an energy converter. The sensing mechanism is actually an information transfer process which requires energy conversion. The energy transfer may be from the object to the sensor or vice versa. A positive output voltage indicates the flow of energy from the object to the sensor. For an example, a thermopile infrared radiation sensor will provide a positive voltage when the infrared flux flows from the object to the sensor i.e. the object is warmer than the sensor. In this case, when the output voltage is zero, both the object and the sensor are at the same temperature. Sometimes a transducer is described as a sensor, but a sensor is not a transducer. The fundamental difference between a sensor and a transducer lies in the fact that the sensor converts any form of energy into electrical energy whereas a transducer converts any form of energy into any other form. A loud speaker is a form of transducer which converts electrical energy into magnetic energy and then to sound energy. Some transducers are called actuators which work through a principle opposite to a sensor. For example, an electric motor is an actuator that converts electrical energy into mechanical energy. A sensor may contain two to three transducers within it. For thermopile example, a chemical sensor consists of two transducers, the first transducer converts the energy related to chemical reaction to heat energy, and the second sensor is a thermopile, which converts the heat energy to an electrical signal. This type of sensor is also called a complex sensor.

1.5 COMMON TYPES OF SENSORS

Considering the different types of energy associated with the signal, the sensors may be (i) mechanical, (ii) electrical, (iii) thermal, (iv) radiant, (v) chemical and (vi) magnetic. It is important to understand the types of measurements associated with the six types of energy. Table 1.2 gives a relation between types of energy and the measurands

Table 1.2 Types of energy and the measurand

Type of energy	Measurand
Mechanical	Length, area, volume, force, pressure, acceleration, torque, mass flow, acoustic intensity, etc
Electrical	Charge, current, voltage, resistance, inductance, capacitance, dielectric constant, polarization, frequency, electrical pole, etc.
Thermal	Heat flow, entropy, temperature
Radiant	Intensity, phase, refractive index, reflectance, transmittance, absorbance, wavelength, polarization, etc
Chemical	Concentration, composition, oxidation/reduction potential pH etc
Magnetic	Field intensity, flux density, permeability., magnetic moment

Considering the physical quantity which is to be measured, the different types of sensors are mentioned in Table 1.3

Table 1.3 Types of sensors considering physical quantity to be measured

Physical phenomenon	Type of sensor
Temperature	Thermocouples, thermistors
Strain on an object	Strain gauge
Weight and load	Load cells
Displacement	LVDT
Acceleration, vibration	Accelerometers
AC or DC current	Current transducers
High voltage potentials	Voltage transformer
Longitudinal and latitudinal position of a place	GPS sensors

1.5.1 Temperature sensors:

The most common temperature sensors are Thermistors, Thermocouples, RTDs etc.

Termistors are resistance thermometers whose resistance is dependent on temperature. A metallic oxide is pressed in the shape of a bead, disk or cylinder which is covered with an epoxy. Different types of thermistors are shown in Fig 1.2.

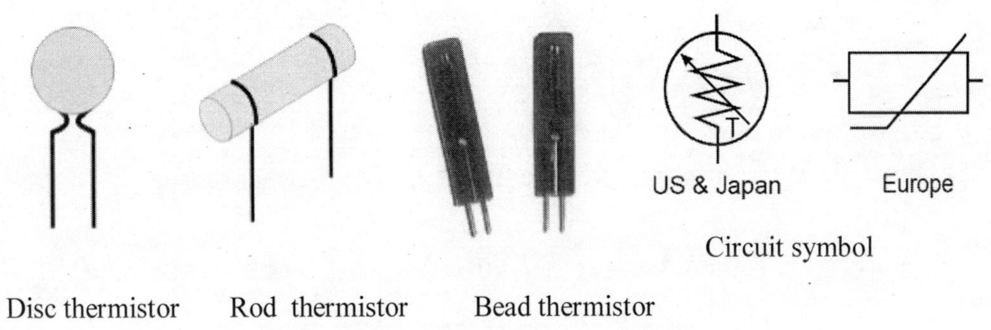

Fig. 1.2: Different types of thermistors
(Source : Internet studentlession.com)

The thermocouple is working on the principle that if two dissimilar metals are in contact with each other at one end, they create a small voltage potential under a change in temperature. The phenomenon is called the "Seebeck effect". Thermocouples are best suited for high temperature measurement. Fig 1.3 shows a picture of a thermocouple.

Fig. 1.3: Thermocouple

RTD (Resistance Temperature Detectors) are more sensitive and precise and are normally used in Pharmaceuticals. They are also costly because of their platinum content. Common types of RTDs are Pt100 and Pt1000 (Pt means Platinum and 1000 means 1000 ohm at 0^0C). Fig 1.4 shows a photograph of an RTD

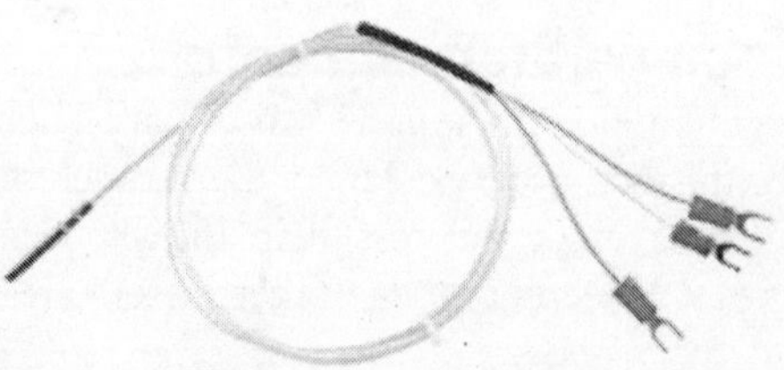

Fig 1.4: Photograph of Resistance temperature detector
(*Source : tradeeasy.com)*

Table 1.4 shows the different aspects of the three types of thermal sensors.

Table 1.4: Aspects of different thermal sensors

	Thermistor	Thermocouple	RTD
Temperature range	-100 to 325^0C	200 to 1750^0C	-200 to 650^0C
Linearity	Exponential	Non-linear	Fairly linear
Power required	Constant voltage or current	Self-powered	Constant voltage or current
Response time	Fast 0.12 to 10 s	Fast 0.1 to 10 s	Generally slow, 1 to 50 s
Accuracy	0.05 to 1.5^0C	0.5 to 5^0C	0.1 to 1^0C
Cost	Low to moderate	Low	High

1.5.2 Strain gauges

Strain is a non-dimensional quantity and is defined as a ratio of change in length to the original length of an object. A positive strain indicates extension and thus represents a tensile force while a negative strain indicates a compressive force. The strain gauge is composed of a metallic foil insulated by a flexible material. The gauge carries a current through two leads. When there is a change in length, there is a change in resistance. The following formula is used to obtain the change in resistance:

$$\frac{\Delta R}{R} = k\,\varepsilon \qquad (1.1)$$

Where, ΔR = change in resistance in ohm
R = Resistance in ohm
k = gage factor
ε = strain

Ex 1.1: In a strain gauge with a resistance of 30 ohm there is a change in resistance of 1 ohm under a loading condition. Obtain the strain if the gage factor is 2.

Answer: The required strain = (1/30)/2 = 0.0167 i.e. 1.67%

Example 1.1

The change in resistance is obtained from the Wheatstone bridge principle as shown in Fig 1.5

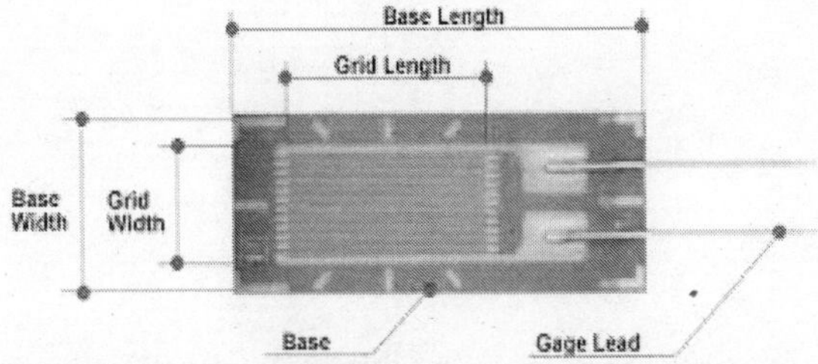

Picture of a strain gauge

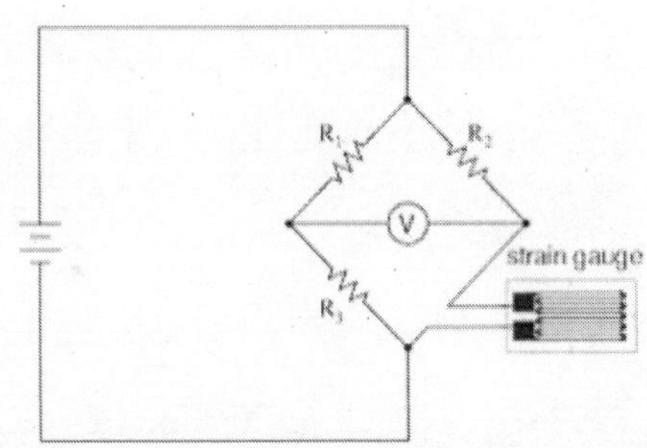

Position of a strain gauge in a Wheatstone bridge

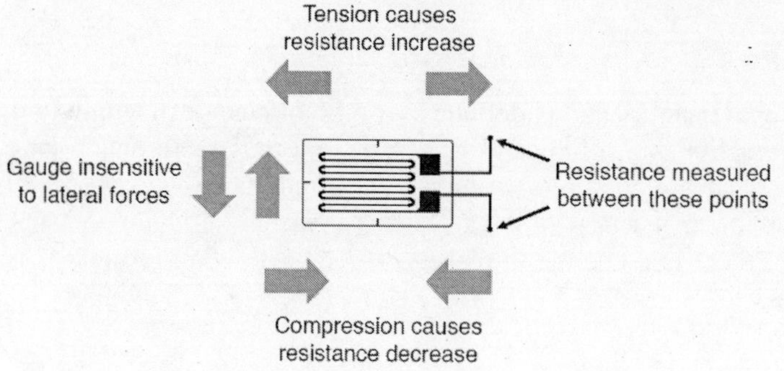

Fig 1.5: Working of a strain gauge

(*Source : allaboutcircuits.com*)

There may be two strain gauges or four strain gauges attached to a Wheatstone bridge. Accordingly, the arrangement is called half or full bridge circuits. The half bridge circuit yields twice the output voltage compared to a single strain gauge, also called a quarter bridge circuit. The full bridge circuit is more sensitive and typically preferred for precision measurement. The position of strain gauges in half bridge circuit and full bridge circuit are shown in Fig 1.6

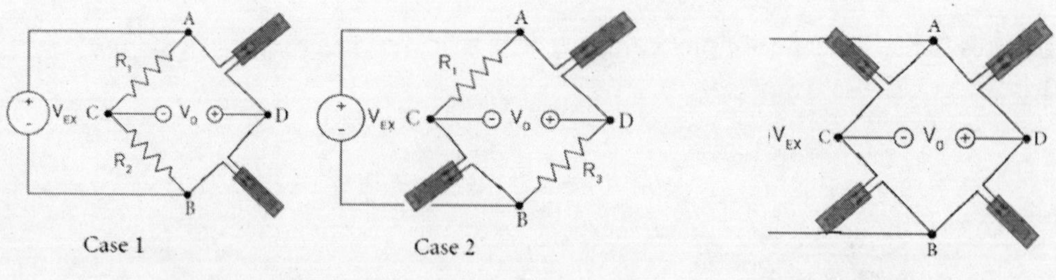

Half Wheatstone bridge Full Wheatstone bridge

Fig. 1.6: Position of two and four strain gauges in a Wheatstone bridge

(*Source: www.bestech.com.au*)

1.5.3 Load cell sensors

A load cell sensor mainly measures the force. It is a transducer which converts the mechanical force into measurable electrical force. The electrical force can be measured through electrical units. The load cell sensors are always bonded along with strain gauges. The strain gauges are made of stainless steel and are fixed inside the load cell using an adhesive. When a force is applied to the load cell, there may be bending, stretching or compression of the strain gauge. As the length and cross-section of the strain gauge change, the resistivity of the strain gauge also changes. A change in resistivity imparts a change in the output voltage. A proper calibration of the output voltage directly gives the value of the external force.

The load cell also works on the principle of the Wheatstone bridge, as shown in Fig. 1.7

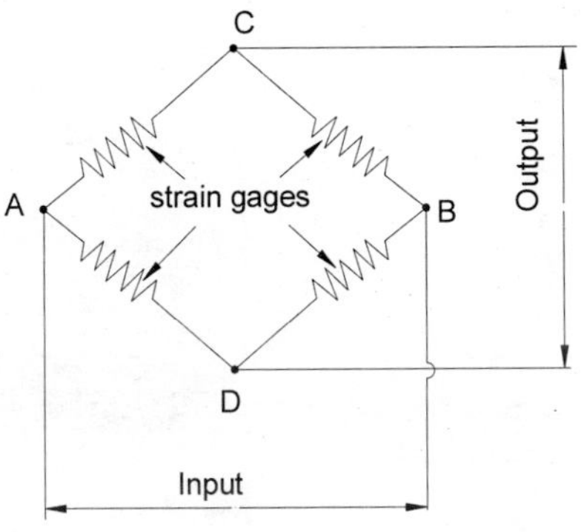

Fig. 1.7: Principle of Wheatstone Bridge

The input voltage is applied at A and B by a signal conditioner or digital display and the output voltage is measured through a digital display. When no load is applied, the bridge is balanced and will not show any input. When a load is applied, the input voltage will change due to a change in resistance between points A and B. This will change the output voltage which can be measured on the digital display unit. Although load cells utilize the functioning of the strain gauges, they cannot be said to be a part of strain gauges. While a strain gauge uses a single resistive element, the load cell utilizes the principle of Wheatstone bridge with four resistors.

There are many types of load cells as discussed below

a) Beam type load cell

This type of load cell is mainly used in tank/vessel weighing and floor scales. The load cell may be single ended or double ended. The single ended load cell is mounted on one end and the load is applied on the other end. The double ended load cell is mounted on both ends and the load is applied at the middle. The double ended load cell is used for higher capacities. Fig 1.8 shows a picture of beam type load cell

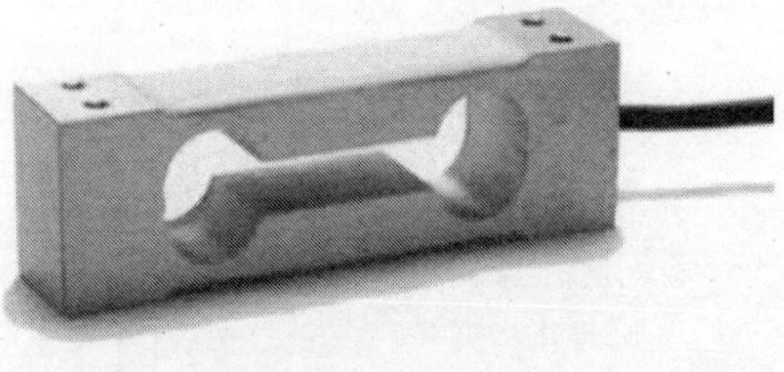

Fig 1.8: Beam type load cell

(Source :transducertechniques.com)

b) S – type load cell

This type of load cell is mainly found in hanging scales or suspended vessel weighing. The load cell can be used in both tension and compression. For a tensile load, the cell is attached to one end in a suspended position and the load is placed on another end. For a compressive load the cell is fixed at two ends and load is applied. Fig 1.9 shows an S-type load cell

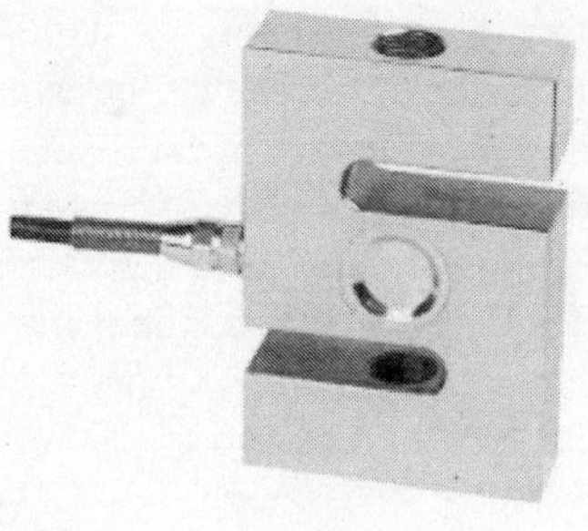

Fig. 1.9: S -type load cell

(*Source: Indiamart.com*)

c) Shear web load cell

These load cells are ideal for verifying force measurement in press applications. The cell can be used for compression when sandwiched between two components or for tension via the threaded hole at the centre. Fig 1.10 shows a picture of the shear web load cell.

Fig. 1.10: Shear web load cell

d) High capacity compression load cell

The load cell consists of columnar element of high strength stainless steel. The sensor uses 16 number of 350ohm resistance strain gauges. Fig 1.11 shows a picture of high compression load cell

Fig 1.11: High capacity compression load cell

(*Source : Encardio-rite*)

e) Single point load cell

This load cell is a low capacity load cell. It is unique because it can measure the load accurately even if the weight is placed at any position on the scale. It is highly used in the industries. Fig 1.12 shows a picture of a single-point load cell

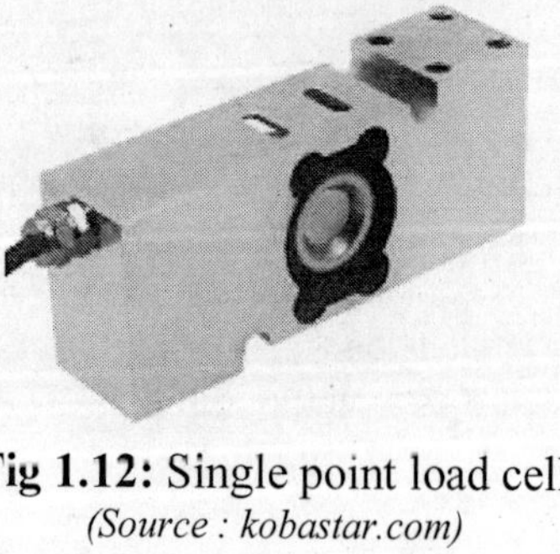

Fig 1.12: Single point load cell

(*Source : kobastar.com*)

1.5.4 LVDT, Linear variable differential transducer

A linear variable differential transducer (LVDT) is commonly used equipment for measuring the element of linear displacement. LVDT is a very important sensor in civil engineering. It is actually named as electromechanical passive inductive transducer. It does functioning due to a combination of electrical and mechanical processes. It is passive because it cannot generate energy. It is inductive because it is operative due to the formation of an electromagnetic induction field. Transducer because it transforms energy from one form to another. When the core moves within the coil there is a change in induced voltage in the secondary coil. This induced voltage is related to the displacement; hence the displacement can be easily measured. Fig 1.13 shows the operation of an LVDT

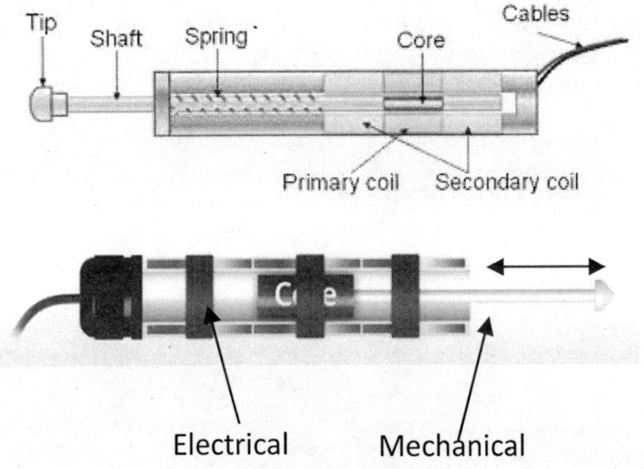

Fig 1.13: Operation of an LVDT

1.5.5 Accelerometer sensor

An accelerometer sensor is a device that measures the acceleration of anybody or an object. It is mainly used on smartphones, wearable devices, etc. The working principle of an accelerometer is simple. It measures the acceleration force in g and takes measurement in one, two or three directions. The most common accelerometers which are used now-a-days are three dimensional accelerometers which consist of three separate accelerometers. Each accelerometer measures acceleration in one of the three directions x, y and z. A three-dimensional accelerometer is positioned in such a way that the x-axis accelerometer sensor points left, the y–axis accelerometer points downwards and the z-axis accelerometer sensor points forward. In such a position, the x-axis will show 0.0g, the y-axis will show 1.0 g and the z-axis will show 0.0g. There are three types of accelerometers, namely

- MEMS capacitive accelerometer
- Piezoelectric accelerometer
- Piezoresistive accelerometer

MEMS capacitive accelerometer – This type of accelerometer is the cheapest and the smallest among the three. MEMS stands for Micro Electro Mechanical System. Fig 1.14 shows the operation of the capacitive accelerometer. When the mass moves due to acceleration, there is a change in the distance between the plates which causes a change in the capacitance of the capacitors.

Piezoresistive accelerometer – the piezoresistive materials are attached to flexible beams. When the beams move there is a change in the resistance of the piezoelectric material, which causes a change in the electrical signal. Fig 1.15 shows the working principle of a piezoresistive accelerometer.

Piezoelectric accelerometer - The piezoelectric accelerometer works in a similar way as the piezoresistive accelerometer works. When the piezoelectric material deforms, there is a change in the electrical signal. Fig 1.16 shows the working of a piezoelectric accelerometer

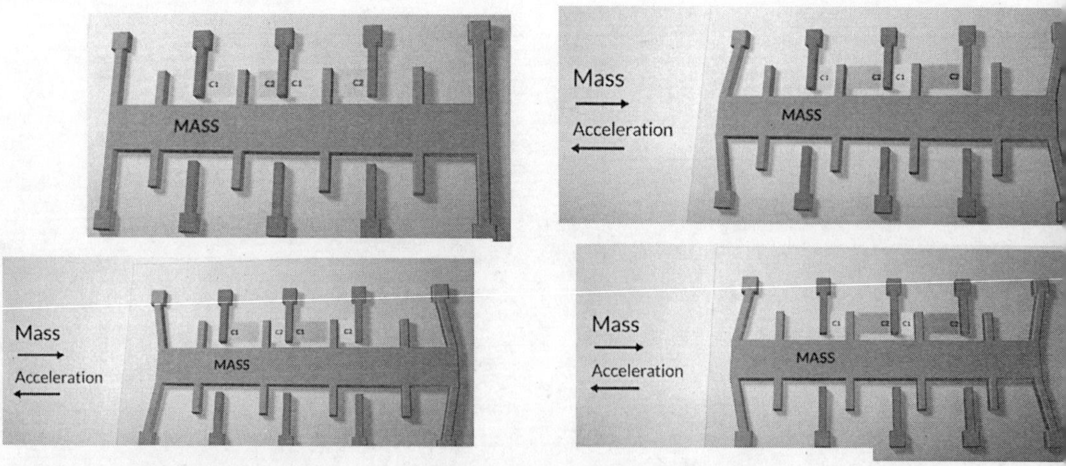

Fig 1.14: Operation of a capacitive accelerometer

Source : You tube Transfer Multosort Elektronik

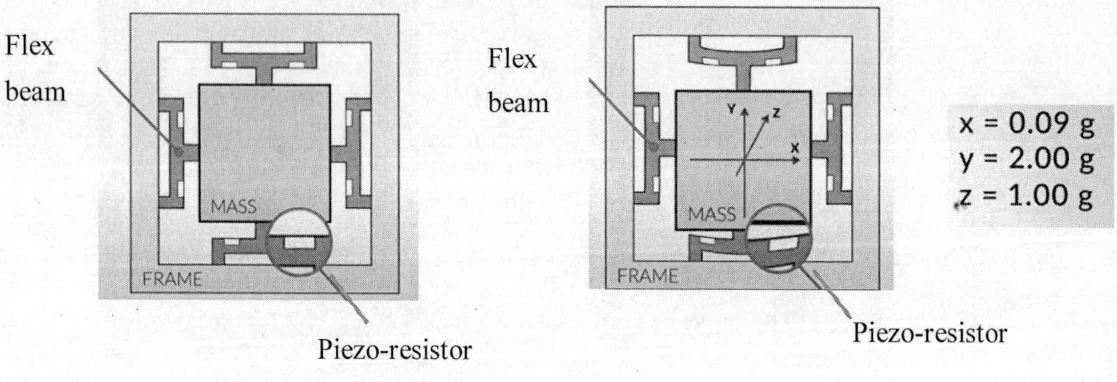

Fig 1.15: Operation of a piezo-resistive accelerometer

Source : You tube Transfer Multosort Elektronik

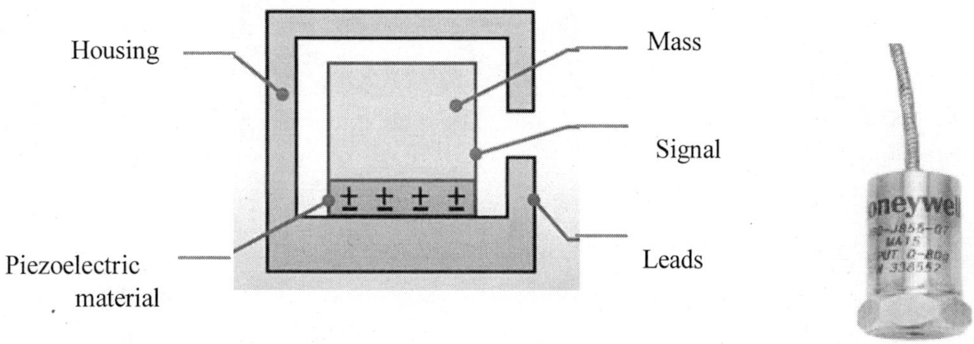

Fig 1.16: Operation of a piezo-electric accelerometer

(Source : directindustory.com)

1.5.6 Current Transducer

Current transducer is a device which converts the AC current into a proportional DC current or voltage.

The current transducer has four parts, namely the sensitive component, conversion component, conversion circuit and power circuit. Fig 1.17 shows picture of current transducers.

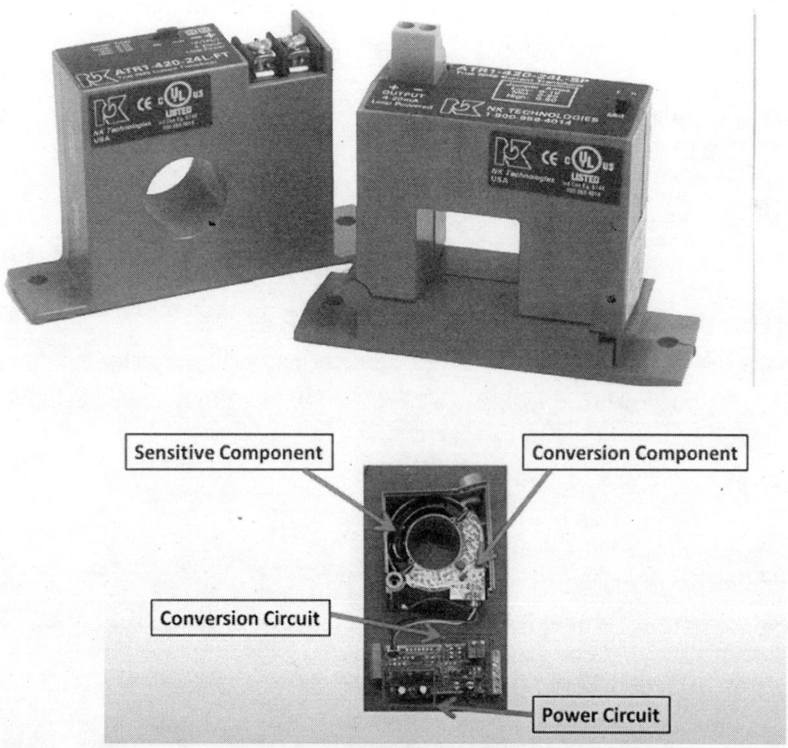

Fig 1.17 Current transducer

(Source : www.set-transducer.com)

The operation of a current transducer is shown in Fig. 1.18. The measured electrical parameter in the form of current, voltage, power or frequency is passed to a sensitive component which sends the parameter to a conversion component. The conversion generates a small current signal which is finally converted to an industry standard electrical signal like voltage, current or Recommended Standard (RS) 485. At the end, the output signal goes to the output terminal as a display, PLC, alarm unit or in automation control. The current transducer has a power circuit which provides power to the conversion component and conversion circuit.

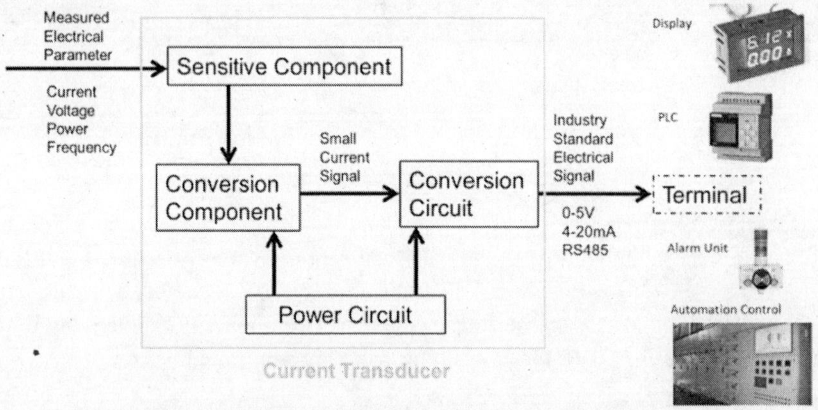

Fig 1.18: Operation of a current transducer

The main functions of the current transducer are as follows:

1. Isolation of input current from output current
2. Conversion function
3. Enhance signal for long distance transfer
4. Safety function

1.5.7 Voltage transducer

A voltage transducer provides a DC current or voltage directly proportional to an AC input voltage. The transducer has an input transformer which isolates the transducer from the voltage input.

Voltage transducers are used to monitor or log voltage levels in AC and DC systems. These transducers provide a fully isolated, 4-20 mA output directly proportional to the DC input voltage. Fig 1.19 shows a picture of the voltage transducer.

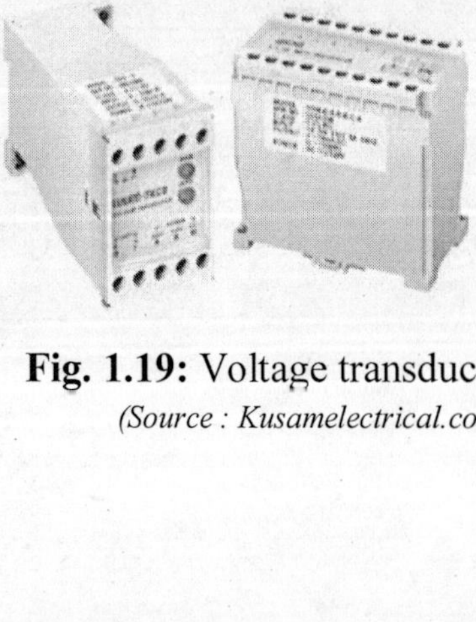

Fig. 1.19: Voltage transducer

(Source : Kusamelectrical.com)

1.5.8 GPS sensors

The GPS sensors are receivers with antennas that use navigation through 24 satellites in orbit around the earth to provide information on position, velocity and time.

1.6 CLASSIFICATION OF SENSORS

There is no single way to classify the sensors. However, they can be broadly classified into two categories namely active and passive. An active sensor requires an external power to operate; few examples are strain gauges (change in resistance indicates a change in strain), thermistors (change in resistance indicates change in temperature), capacitive or inductive sensors (change in capacity or inductance indicates change in position), etc. The passive sensors do not require any external power, rather they are self powered sensors. Examples are solar cells, microphones, piezoelectric sensors, thermoelectric sensors, etc.

Another way to classify the sensors is contact and noncontact sensors. The contact sensors will need to have contact with the measuring substance, like a thermistor sensor needs physical contact to measure the temperature of a metal surface. However, the same sensor may also be a non-contact sensor when it is used to measure the ambient temperature of a car.

Again, the sensors may be absolute or relative. The absolute sensors measure the absolute value of the measurable quantity, for example, a thermistor measures the temperature of a substance and that temperature is the absolute value. A relative sensor measures the relative value. A thermocouple measures the temperature difference between two junctions. Another relative sensor is a pressure sensor. The pressure sensor measures the additional pressure over an existing atmospheric pressure.

However, in a broad sense the sensors are classified as shown in Table 1.5

Table 1.5: Broad classification of sensors

Broad area	Detectable area	Measured output	Physical laws	Specifi-cations	Applications	Other types
Sensors	• Electric • Magnetic • Electro--magnetic • Acoustic • Chemical • Optical • Thermal • Mechanical	• Resistive • Capacitive • Inductive • Current • Voltage • Resonant • Optical • Mechanical	• Electro-resistive • Electro-chemical • Electro-optic • Magneto-electric • Magneto-caloric • Magneto-resistive • Photo-electric • Photo-elastic	• Accuracy • Sensitivity • Stability • Response time • Hysteresis • Frequency response • Resolution	• Consumer products • Military applications • Infrastructur e • Energy • Heat / thermal	• Power • Interfaces • Structure

Broad area	Detectable area	Measured output	Physical laws	Specifi-cations	Applications	Other types
	• Radiation • Biological		• Photo-magnetic • Photo-conductive • Thermo-magnetic • Thermo-elastic • Thermo-optic • Thermo-electric	• Linearity • Hardness • Cost • Size • Weight • Construction material • Operating temperature	• Manufacturing • Transportation • Automotive • Avionics • Marine • Space • Scientific	

1.7 TERMS USED IN SENSOR APPLICATIONS

There are several terms which are used in the sensor applications. Some of the common terms are defined as follows :

Sensor : A device which provides a usable output in response to a specified measurand. It is a device that receives and responds to a physical stimulus and transmits a resulting stimulus.

Instrument– It is a device that provides a quantity with its present value under observation.

Accuracy – degree of closeness of a measured value of a quantity to the expected value.

Measurement – The process of determining the amount, degree, capacity by comparison (direct or indirect) with the accepted standards of the system units being used.

Resolution – The smallest change in the measured quantity.

Precision – It means the repeatability of a measurement i.e., successive readings do not differ.

Precision can be expressed through the following formulation:

$$Precision = 1 - \left|\frac{x_n - \bar{x}}{\bar{x}}\right| \quad (1.2)$$

Where x_n = measured value

$$\bar{x} = average\ value\ or\ expected\ value$$

Expected value – The design value or the most probable value expected from a measurement.

Hysteresis –It is a lag between input and output in a system upon a change in direction. Hysteresis is something that happens with magnetic materials so that, if a varying magnetizing signal is applied, the resulting magnetism that is created follows the applied signal, but with a delay.

Dead Zone/band – The range of possible values for an instrument without any reading although there are changes in the measured parameter.

Nominal value–provided by the manufacturer in the user manual for some input and output values.

Bias – A constant error that occurs to the instrument when the pointer is not starting from a zero scale.

Range – Minimum and maximum values for smooth functioning of an instrument stated by the instrument manufacturer.

Sensitivity – The ratio of the change in output (response) of the instrument to a change in input.

Transducer – A device that is actuated by power from one system and supplies power in usually another form to a second system. (Webster's New Collegiate Dictionary, 1998)

Actuator – A mechanism for moving or controlling something indirectly instead of by hand (Webster's New Collegiate Dictionary, 1998)

Decibel - Measuring scale to measure a parameter which varies over an extensive range. For example, the human eye can see a luminance over a range 10^{-6} cd/m^2 to 10^6 cd/m^2. Thus, it is not easy to measure a luminance of 10^3 cd/m^2. It is obtained through the following steps:

a) Select the threshold value of a measuring parameter. For the human eye, the threshold value may be 10^{-6} cd/m^2.
b) Divide the given quantity by the threshold value.
c) Take the logarithm of base 10 for the ratio as obtained in (b)
d) Multiply the logarithmic value by ten if it is power related like power, energy, power related, etc. Otherwise multiply the logarithmic value by 20 for field quantities like voltage, current, force, pressure etc.

The luminance of 10^3 cd/m^2 can be expressed through dB (decibel) as follows:

Luminance $= 10 \log_{10}(10^3/10^{-6}) = 90$ dB

Thus, the human eye can see a luminance from 0 to 120 dB.

Example 1.2

Ex. A power sensor is rated to measure power -30 dBm to 20 dBm. Calculate the range in terms of power considering the reference value of the power = 1 mW

Ans : Let P_0 and P_m be the range of power in mW

For lowest value $-30 = 10 \log_{10}(P_0/1)$

$P_0 = 0.001$mW

For highest value $20 = 10 \log_{10}(P_m/1)$

$P_m = 100$mW

Thus the range of the sensor = 0.001 mW to 100 mW

Absolute Zero: Temperature at which thermal energy is at a minimum. Defined as 0 Kelvin, calculated to be –273.15°C or –459.67°F.

Calibration: The process of adjusting an instrument or compiling a deviation chart to correlate its reading to the actual value being measured.

Damping: The reduction of vibratory movement through dissipation of energy. Types include viscous, coulomb, and solid.

Thomson Effect: When current flows through a conductor within a thermal gradient, a reversible absorption or evolution of heat will occur in the conductor at the gradient boundaries.

Analog sensor – This type of sensor has an output signal that is continuous in magnitude and time.

Deflection instrument – An instrument whose output deflects from an initial position under the influence of a measurand. The deflection is normally proportional to the influence.

Digital sensor – These sensors have a discontinuous output in magnitude and / or time.

Input impedance–The impedance measured across the input terminals of a device

Measurand - A physical quantity whose property or condition is to be measured.

Null instrument – A device which balances the influence of a measurand and shows no deflection.

Readout –this is the display of a measuring device.

Resolution – the lowest change in the measurand value detectable by a device

Difference between sensor and transducer: A sensor converts any type of energy into electrical energy whereas a transducer converts one type to another type. A loudspeaker is an example of a transducer which converts an electrical signal into a variable magnetic field and thereby into sound energy. However, when the loudspeaker is connected to an amplifier, it becomes a microphone. The microphone is an acoustical sensor

1.8 INTERPRETATION OF SIGNALS FROM SENSORS

The output of a sensor is usually an electrical signal which can be channelled, amplified and modified by electronic devices. The output signal may be in the form of voltage, current or charge which can be designated in amplitude, frequency or digital code. Electrical signals from the sensors are fed into a multiplexer (MUX) which is a switch or a gate. The function of the multiplexer is to connect every sensor to an analog-to-digital converter. A block diagram in Fig 1.20 shows a detailed interpretation of signals through a multiplexer.

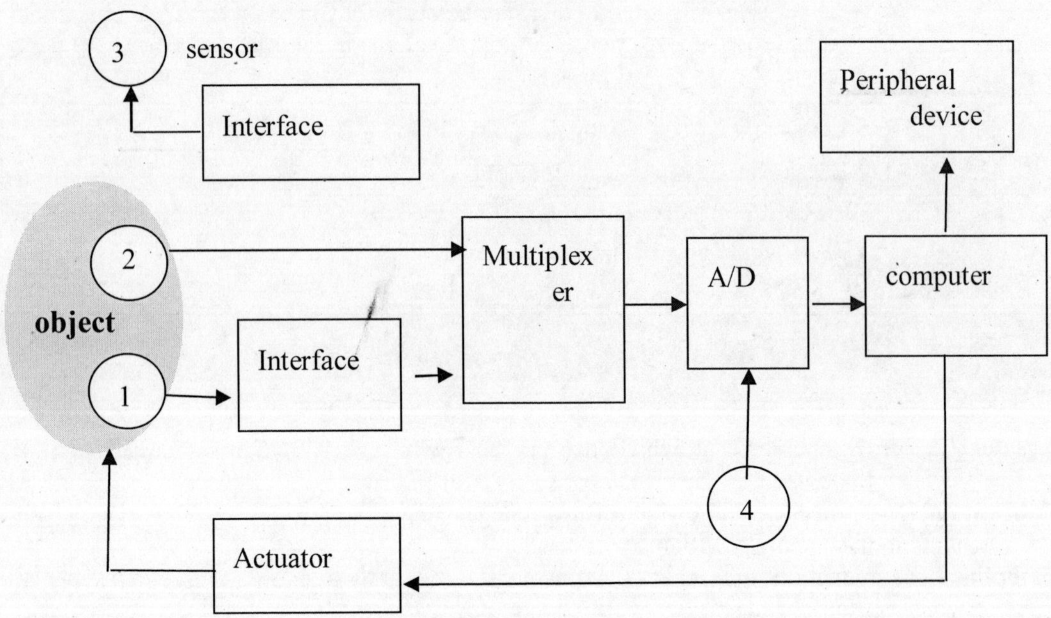

Fig. 1.20: Position of sensors : 1 and 2 contact sensors, 3 and 4 non-contact sensors. The signal from a sensor is sent to the computer through a multiplexer

An example of the interpretation of a signal from a sensor may be mentioned through the opening of a car door. Every door of a car is connected to a sensor in the form of a simple electric switch. The sensor signal goes to car's internal processor since the switch is a digital form, either one or zero. The processor identifies which door is open and sends the signal to a peripheral device as an alarm. The driver acts as an actuator and takes the necessary step (closes the door).

1.9 TYPES OF INSTRUMENTATION

Instruments can be divided into different types based on different criteria.

Based on the method of measurement, the instruments may be of two types:

a) Null instruments
b) Deflection instruments

A null instrument usually balances the measurement. The null instrument exerts an influence on the measurand and both the influence and measurand become equal but opposite resulting in null measurement. A null instrument thus balances the unknown input against a known standard input. Therefore, the main advantage of the null instrument is that there is very little interaction between measuring system and the measurement. The main disadvantage of the null instrument is that the balancing operation is iterative type and requires more time than a simple measuring sensor. Fig 1.21 shows an example of a null instrument

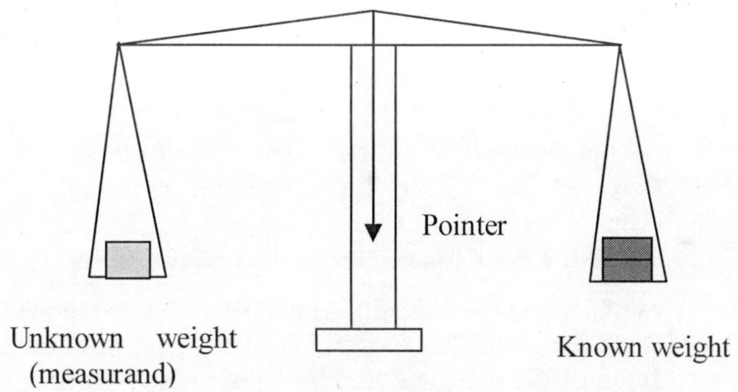

Fig. 1.21: An equal arm balance scale

An equal arm balance scale is a good example of a mechanical manual null instrument. An unknown weight i.e. measurand is placed on one side and a set of known weights is placed on the other side. The known weights are added or removed iteratively so the pointer shows a zero reading i.e. the pointer does not show any deflection. The unknown weight or measurand is then said to have a value equal to the known weight. Other common examples of null instruments are bridge circuits used in load cells, temperature compensated transducers, voltage balancing potentiometer, etc.

A null instrument needs two inputs, one is the measurand and the other is the balancing weights. A comparator compares both the inputs and a non-zero output indicates an iterative feedback. Repeated corrections lead to a parity between the two inputs when the indicator shows a null deflection. Fig 1.22 shows the operation of a null instrument

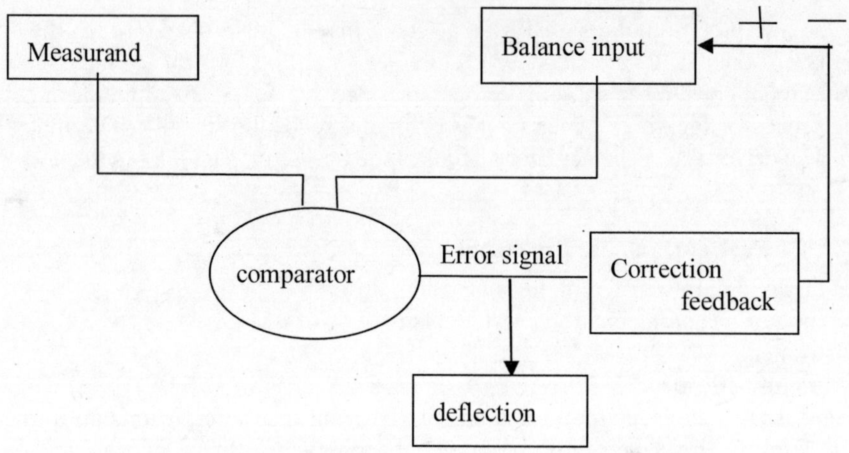

Fig. 1.22: Operation of a null instrument

Deflection instrument uses the deflection from an equilibrium state when the measurand influences the instrument. The deflection from the initial position gives an output reading which is proportional to the influence of the measurand. The deflection instruments are the most common type of instruments which provide a direct measurement without any balancing mechanism or comparator circuit. The proportional response is manipulated through signal conditioning circuits between the prime element and the output scale so that the output reading directly measures the measurand. The main advantage of the deflection instrument is that it can be used for both static and dynamic conditions. The instrument can achieve a high dynamic response. The only disadvantage of this type of instrument is a loss of energy during the deflection operation. This gives an error in the measured value, which is to be accepted. This involves a careful look into the input impedance of the instrument for the output measurement.

A spring scale is a good example of a deflection type instrument as shown in fig 1.23. When a measurand is placed over the plate, the spring deforms. The equilibrium condition will reach when the downward force of the weight W i.e. the measurand is equal to the upward restoring force kx of the spring. A cross linkage or a pointer may be attached to the spring. The pointer position may be mapped out on a scale which is the readout scale. For the input weight W, the read-out scale will give a deflection $x = W/k$. The read-out scale is normally pre-calibrated so that the deflection x directly gives the output measurement as W.

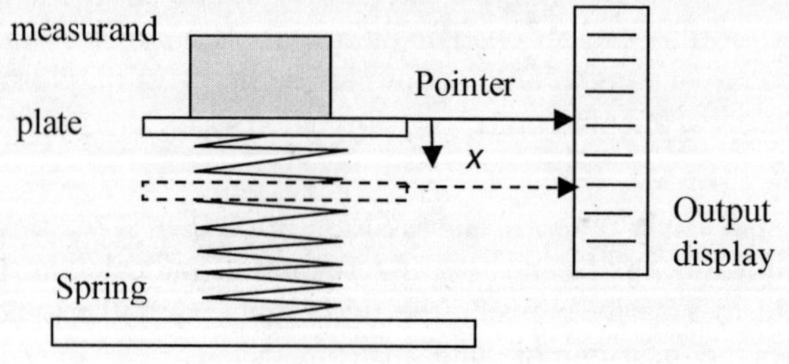

Fig. 1.23: Example of spring plate deflection instrument

The flow diagram logic for the deflection instrument is relatively simple. The input signal is sensed by the instrument which will be depicted through a deflection from the initial setting. The deflection signal is transmitted to a signal conditioner which shows the output reading on a scale. The scale is pre-calibrated with a multiplier to the measured deflection. The output reading will therefore, give a direct measurement to the input signal. Fig 1.24 shows a flow chart of a deflection type instrument.

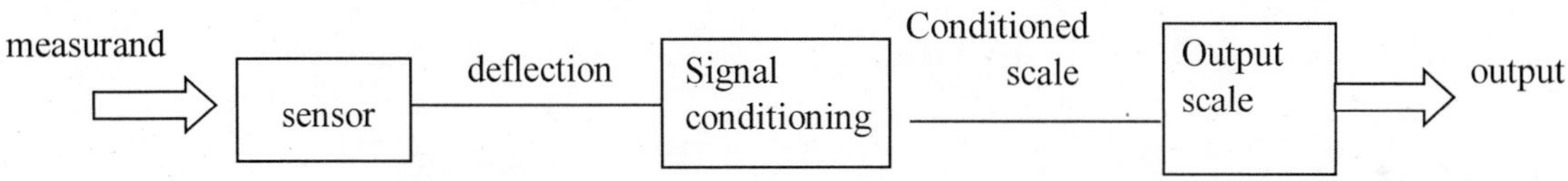

Fig 1.24: Flow chart of deflection instrument

Depending upon the consumption of power, an instrument can be divided into two types

1. Active
2. Passive

An active instrument requires some external power either a battery or an external voltage source for its working. Fig 1.25 shows the operation of an active instrument, where an external voltage is applied. A passive instrument does not need any external power for its operation. For the passive instruments, the output is completely produced by the measurand. Examples of active instruments are transistors, LED lights, IC chips etc. The common passive instruments are resistors, capacitors, potentiometers etc. Fig.1.26 shows a simple example of a passive instrument. When the fluid pressure increases the piston moves up and the pointer gives a reading of the applied pressure. Thus, the pressure gauge does not require any external power and the output is produced by the pressure of the fluid only.

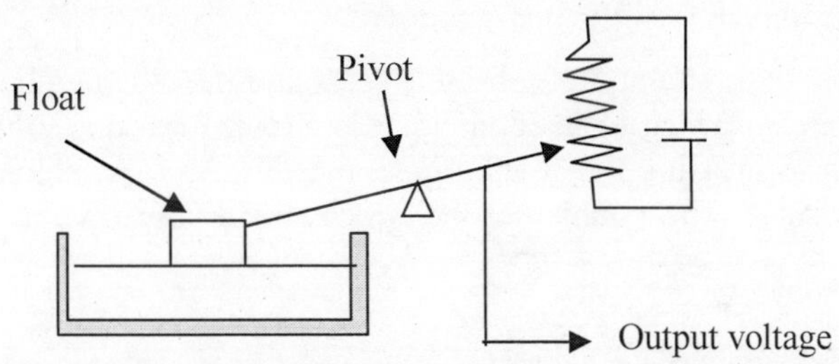

Fig 1.25: Petrol tank level indicator – an example of an active instrument

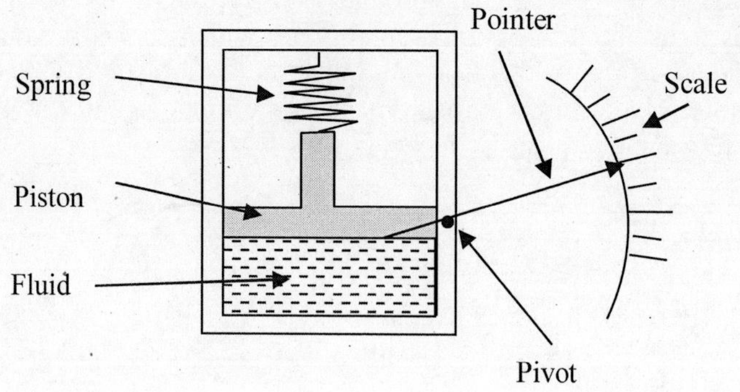

Fig. 1.26: Pressure gauge – an example of a passive instrument

The petrol tank level indicator is an example of the active instrument in which an external battery is required to change the output voltage. The pressure gauge is an example of a passive instrument, since it does not need an external power to show the output.

Depending upon the continuity of the output the instrument may be of two types:

1. Analog
2. Digital

An analog instrument gives a **continuous** output as the quantity being measured changes. Thus, the analog instrument gives an infinite number of continuous outputs within the range of the instrument. A deflection type instrument is a good example of an analog instrument. As the input value changes the output of the instrument gives an infinite number of continuous readings over the range of the instrument.

A digital instrument gives an output at discrete steps and thus the number of readings are limited. Fig 1.27 shows a cam which opens and closes a circuit for a one time during one revolution. The instrument is to measure the revolutions, and thus, the switching operation is counted by an electronic counter. The counter cannot differentiate any revolution or count any incomplete revolution.

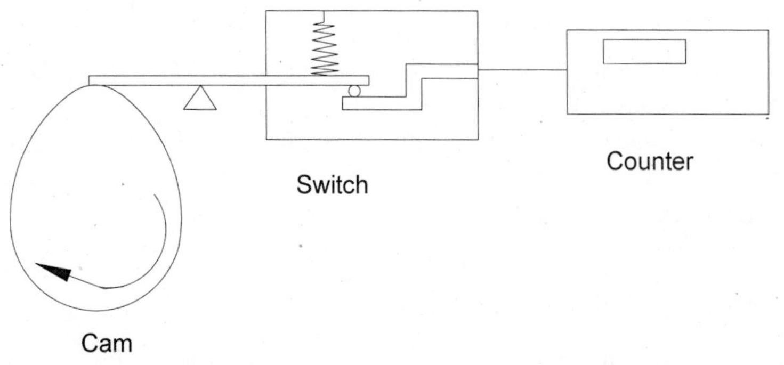

Fig. 1.27: Revolution counter – an example of a digital instrument

Depending upon the inclusion of a microprocessor the instruments may be of two types :

1. Smart instrument
2. Non-smart instrument

A smart instrument contains a microprocessor, whereas, a non-smart instrument does not contain any microprocessor

1.10 SENSOR SPECIFICATIONS

The performance of any sensor depends upon an accurate prescription of different components, performance of these components, adjustment or modifications of the components, monitoring of the components, diagnosis of the malfunctioning of the components, etc. All these functional requirements can be clubbed into specifications of the sensor. The performance of a sensor depends upon the three types of performance measures:

1. Speed of performance
2. Stability and
3. Accuracy.

For performance specification in the analytical domain, two types of dynamic models are used :

1. Differential equation models in the time domain
2. Transfer – function models in the frequency domain

In the time domain, the performance parameters such as riser time, peak time, settling time, and % overshot may be specified. In the frequency domain, parameters such as bandwidth, static gain, resonant frequency, magnitude at resonance, impedances, gain margin and phase margin may be specified

1.11 INSTALLATION OF SENSORS

The sensors are installed either permanently or temporarily. The permanent installation means that the position of the sensors will not be changed during the lifetime of the machine. In contrast, the temporary installation means an option to change the position of the sensor.

1.11.1 Permanent installation

For critical machines like turbines, fans, motors, coal processing equipment, gearboxes, blowers, etc the sensors are to be installed permanently. Different vital departments like oil and gas, aviation, rail, pharmaceuticals, communications, finance, marine sectors, etc need permanent installation of sensors to maintain the constant service with a minimum maintenance. Permanent installation is also required for inaccessible machines.

1.11.2 Temporary installation

When sensors are installed temporarily on the outside, following vulnerable positions are to be avoided :

a) positions exposed to direct sunlight
b) facades
c) ventilation shaft
d) sunshades
e) walls in front of kitchen chimney
f) exposed positions of balconies
g) inaccessible areas
h) moving supports
i) Dark surfaces
j) Adjacent to exhaust fans
k) Interference from trees and buildings

In most of the important and sensitive civil construction sensors are installed permanently for a continuous monitoring of the performance of the system. An underground metro station is an important and sensitive civil construction. The different sensors which are usually installed in a metro station are shown in Fig. 1.28

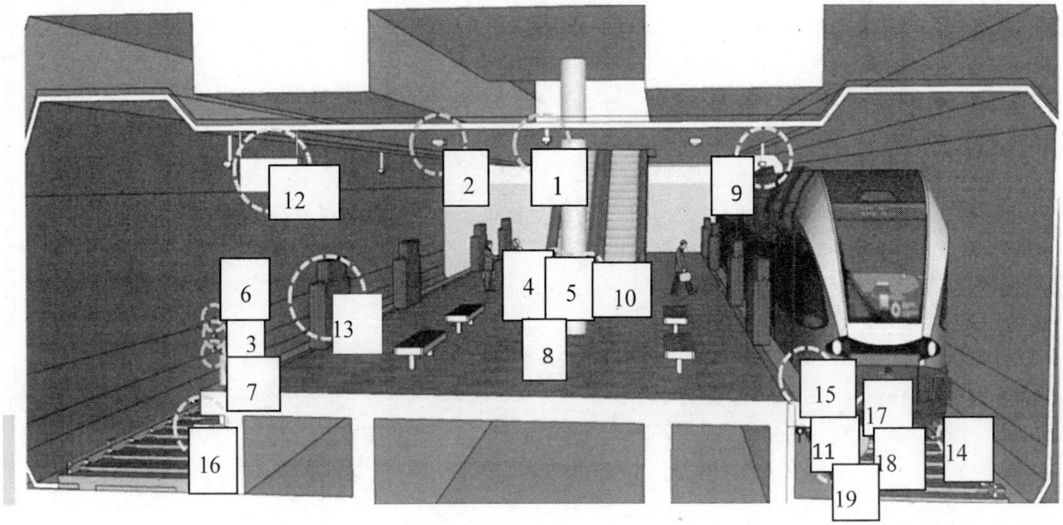

Fig 1.28: Rail transit sensor installation (After Shen et al. 2022)

1. HD camera
2. Infra-red smoke sensor
3. Gas sensor
4. Temperature and Humidity sensor
5. Wind speed sensor
6. Upper flood sensor
7. Lower floor sensor
8. Evacuation road map
9. Drop test camera
10. Emergency stop button device
11. Environment sensor
12. Shut down indicator lamp
13. Platform door
14. Target to auto tracking base station
15. Auto tracking base station
16. Tiltometer
17. Track material monitoring
18. Orbit displacement
19. Ballast subsidence monitor

(Source : https://www.mdpi.com/1424-8220/22/9/3215/pdf?version=1650805796)

Similarly, different sensors which are normally installed in an elevated bridge and a tunnel are shown in Fig 1.29 and 1.30 respectively.

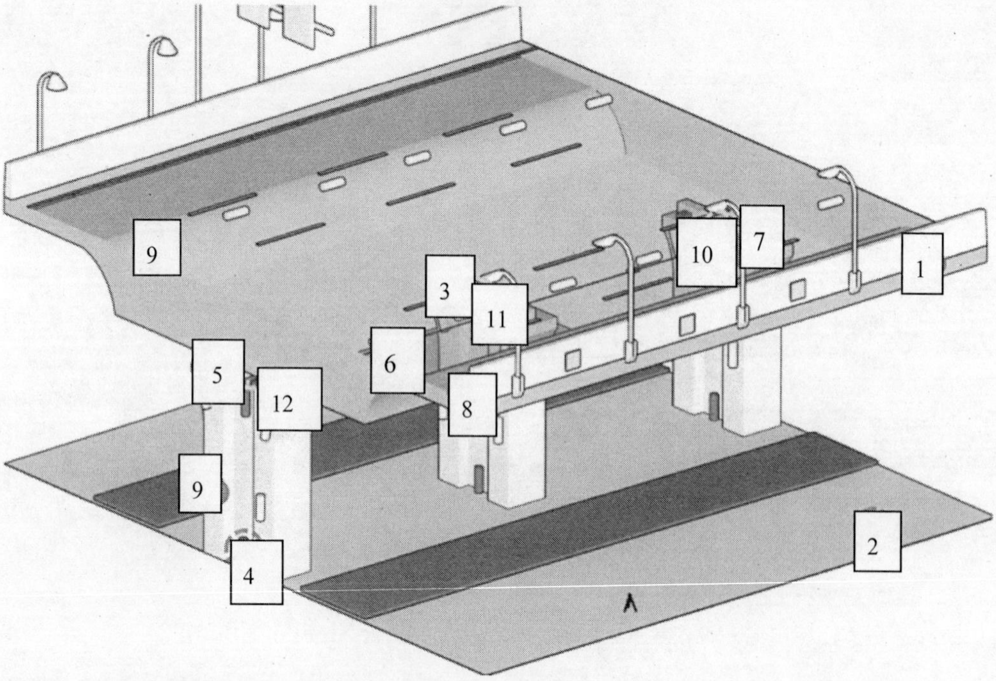

Fig. 1.29: Sensor installations in an elevated bridge (After Shen et al. 2022)

1. Fencing sensor
2. Camera
3. Extensiometer
4. Titometer
5. Automatic tiltometer
6. Measurement sensor
7. Data recorder
8. Strain sensor
9. Elastic strain gauge
10. Frequency sensor
11. Three axis accelerometer
12. Strain gauge

(Source : https://www.mdpi.com/1424-8220/22/9/3215/pdf?version=1650805796)

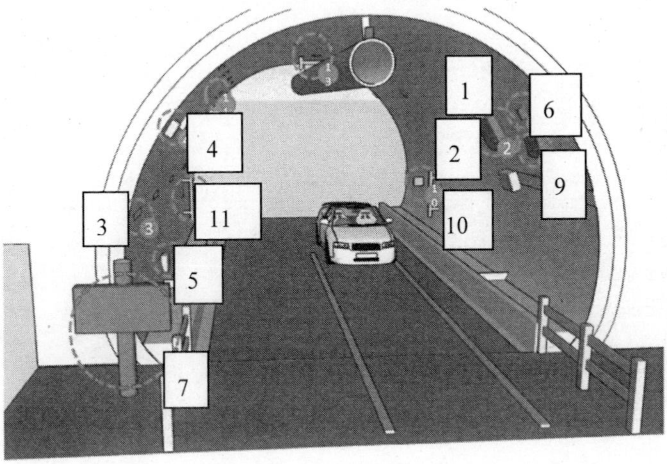

1. Crack displacement gauge
2. Remote wireless unit
3. Crack detection accelerometer
4. Crack detection adapter
5. Data recorder
6. Strain gauge
7. Warning light
8. Network terminal
9. Fibre optic sensor
10. Digital camera
11. Type I basic size

Fig. 1.30: Sensor installation in an underground tunnel (After Shen et al. 2022)

(Source: https://www.mdpi.com/1424-8220/22/9/3215/pdf?version=1650805796)

UNIT SUMMARY

What have we learnt from module 1 ?

Module 1 is an introduction to the main course on Instrumentation for civil engineering applications. Following topics have been covered in this module :

1. Basic definitions of measurement, instrumentation and sensor.
2. Brief discussions on different types of sensors with figures.
3. Basic difference between a sensor and a transducer.
4. Different terminologies used in instrumentation.
5. Different types of instruments.
6. Interpretation of signal from a sensor.
7. Installation of different sensors in different civil structures namely metro station, elevated bridge and an underground tunnel.

EXERCISES

Multiple Choice Questions

1. The task of measurement does not contain the following terms:
   a) Estimation
   b) Calibration
   c) Standard unit
   d) Uncertainty
2. Mechanical measurement is a
   a) Comparison with standard object
   b) Comparison with a similar object
   c) Comparison with electrical output
   d) Comparison with
3. Following measurand does not use mechanical energy
   a) Length
   b) Wavelength
   c) Force
   d) Pressure

4. An LVDT measures
   a) Acceleration
   b) Velocity
   c) Displacement
   d) Time
5. The following is not a thermal sensor
   a) Thermistor
   b) Thermocouple
   c) Resistance Temperature Detectors
   d) Thermometer
6. To measure a temperature of more than 1000^0C, the following thermal sensor is used
   a) Thermistor
   b) Thermocouple
   c) Resistance Temperature Detectors
   d) Thermometer
7. Following sentence is not appropriate for a strain gauge
   a) It measures linear strain
   b) It measures both positive and negative strains
   c) It works on wheatstone bridge principle
   d) Strain has linear dimension.
8. Following sentence is not appropriate for a sensor
   a) Sensor is an energy converter.
   b) Sensor converts any form of energy to the same/other form of energy
   c) Sensor gives a measurable output voltage signal
   d) Sensor is not a transducer.
9. Number of strain gauges attached to a wheatstone bridge may be
   a) 1, 2 or 3
   b) 2, 3 or 4
   c) 1, 2 or 4
   d) 1, 3 or 4
10. Following type of load cell is not available:
    a) Beam type load cell
    b) S-type load cell
    c) Shear wave load cell
    d) Mass and spring type load cell
11. Following type of accelerometer is not available:
    a) MEMS capacitive accelerometer.
    b) Piezoelectric accelerometer.
    c) Piezoresistive accelerometer.

d) Spring type accelerometer.

Answers of Multiple Choice Questions

1(b), 2(a), 3(b), 4(c), 5(d), 6(b), 7(c), 8(b), 9(c), 10(d), 11(d)

SHORT AND LONG ANSWER TYPE QUESTIONS

1. Define (a) measurement (b) instrumentation.
2. What are the different methods of measurement ?
3. What are the different physical variables used in instrumentation ?
4. What are different types of instruments based on physical variables ?
5. Define sensor.
6. How sensors are selected for a given work.
7. What are different types of sensors ?
8. Write short notes on
   (a) temperature sensor
   (b) accelerometer sensor
   (c) strain gauge
   (d) load cell sensor
   (e) LVDT
9. How sensors are classified according to
   (a) Measured output
   (b) Physical laws
   (c) Applications
10. Define following terms: sensitivity, hysteresis, precision, expected value, range, resolution
11. Write a short note on the interpretation of signals from sensors.
12. Differentiate between null instrument and deflection instrument
13. Differentiate between active instrument and passive instrument.
14. Write short note on Analog instrument and digital instrument
15. What are the different types of sensors installed in an elevated bridge ?
16. Name different types of sensors installed in an underground tunnel.
17. Write a short note on different sensors installed in an underground metro station.

ANNEXURE - I

MEASUREMENT ERRORS:

If ideal conditions are applied the measured value of a parameter will be the true value. However, it is almost impossible to obtain a true value of any measurement. Sometimes, the average value of an infinite number of measurements is also considered as a true value assuming that the summation of negative deviations cancels the summation of positive deviations. The difference between the measured value and the true value is called the error. A brief discussion on error is mentioned below.

Types of error :

There are three types of errors:

1. Instrumental error
2. Observational error
3. Environmental error.

Instrumental error is due to the manufacturing defects of the instrument, improper use of the instrument , improper calibration, inherent errors, etc. The observational error is self explanatory. Environmental errors are due to changes in temperature, humidity, environmental pollution, external electrical and magnetic fields, etc.

Calculation of error

Numerically two types of error can be calculated. They are

1. Absolute error
2. Relative error

The absolute error is defined as E_a = mod $\left| V_m - Ve \right|$ where, V_m is measured value and V_e is the exact value.

The relative error is the ratio between absolute errorand the exact value. The relative error is normally expressed in percentage.

Thus, relative error = $\frac{Absolute\ error}{Exact\ value}$

Ex. The measured chain length is 29.56 m whereas the exact value of the chain is 30.0 m. Calculate the absolute error, relative error and percentage of error

Ans : Given, measured value, $V_m = 29.56$ m

Exact value, $V_e = 30.0$ m

Absolute error = mod | 29.56 – 30.0 | = 0.44 m

Relative error = 0.44/30.0 = 0.0147.

Percentage error = 0.0147 x 100% = 1.47%

Example 1.3

Arithmetic mean value :

When a correct measurement is to be obtained, the measurement is taken for a number of times, say n times. If the measurements are $A_1, A_2, A_3, A_4, \ldots A_n$, then

Arithmetic mean $A_{mean} = \frac{A_1+A_2+A_3+A_4+\cdots+A_n}{n}$

The arithmetic mean is very close to the exact or most accurate value.

Absolute error

As defined before, the absolute error is the difference between the true value and the measured value, the true value of a quantity is unknown. However, the arithmetic mean may be considered as true value. Thus absolute error for each measurement can be easily obtained. The absolute error for all the measurements can be summarized as below:

Absolute error for measurement 1 : E_{a1}= mod $|A_{mean} - A_1|$

Absolute error for measurement 2 : E_{a2}= mod $|A_{mean} - A_2|$

Absolute error for measurement 3 : E_{a3}= mod $|A_{mean} - A_3|$

… … … …

… … … …

Absolute error for measurement n : E_{an}= mod $|A_{mean} - A_n|$

The arithmetic mean of all the absolute error is called the mean absolute error, $E_{a\ mean}$

Relative error

The relative error is defined as the absolute error divided by the true value. Hence, the relative error for a number of n observations may be expressed as :

Relative error , $R_{rel} = \frac{Mean\ absolute\ error}{Arithmetic\ mean} = E_{a\ mean}/A_{mean}$

The relative error may be expressed as a percentage, called percentage error.

The percentage error may be written as R_{rel} x 100%

Example 1.4

Ex. In order to obtain the true specific gravity of a soil sample, tests were conducted ten times and the following results were obtained.

2.60, 2.62, 2.615, 2.61, 2.623, 2.622, 2.63, 2.627, 2.61, 2.616

Obtain the true specific gravity, mean absolute error, relative error and percentage relative error.

Also mention how the specific gravity may be expressed.

Ans :

True value of specific gravity = Arithmetic mean of measured values

$$= \frac{2.60 + 2.62 + 2.615 + 2.61 + 2.623 + 2.622 + 2.63 + 2.627 + 2.61 + 2.616}{10}$$

= 2.617

Calculation of mean absolute error :

Observation no	True value	Measured value	Absolute error
1	2.617	2.60	0.017
2	2.617	2.62	0.003
3	2.617	2.615	0.002
4	2.617	2.61	0.007
5	2.617	2.623	0.006
6	2.617	2.622	0.005
7	2.617	2.63	0.013
8	2.617	2.627	0.01
9	2.617	2.61	0.007
10	2.617	2.616	0.001
	total		0.071

Mean absolute error = 0.071/10 = 0.007

Relative error = 0.007/2.617 = 0.002675

% error = 0.002675 x 100% = 0.2675%

The specific gravity of the soil is expressed as = 2.617 ± 0.002

REFERENCES AND SUGGESTED READINGS

1. Alan S Morris (2001). Measurement and Instrumentation Principles, 3rd/ed, Butterworth Heinemann.
2. David A. Bell (2007). Electronic Instrumentation and Measurements 2nd /ed, Oxford Press.
3. S. Tumanski (2006). Principle of Electrical Measurement, Taylor & Francis
4. Ilya Gertsbakh (2010). Measurement Theory for Engineers, Springer.
5. Internet sources
6. Shen, Z., Teng, X., Zhang, Y., Fang, G., & Xu, W. (2022). Guidelines for Installation of Sensors in Smart Sensing Platforms in Underground Spaces. Sensors, 22(9), 3215.

KNOW MORE

For more information related to this topic scan the QR code.

OR

Type this link in your browser

https://www.youtube.com/watch?v=qbKnW42ZM5c

bitly

2 Sensor Installation and Operation

UNIT SPECIFICS

Through this unit we have discussed the following aspects:

- *Prediction of response of sensors to various inputs;*
- *Concept of designing instruments for measuring different physical quantities;*
- *Proper installation of sensors;*
- *Types of sensors and their applications;*
- *Design of sensors for different physical inputs;*
- *Calculate measurement uncertainty.*

The practical applications of the topics are discussed for generating further curiosity and creativity as well as improving problem solving capacity.

Besides giving a large number of multiple choice questions as well as questions of short and long answer types marked in two categories following lower and higher order of Bloom's taxonomy, assignments through a number of numerical problems, a list of references and suggested readings are given in the unit so that one can go through them for practice.

RATIONALE

This fundamental unit on sensor installation and operation helps students to get a primary idea about the essential operation of sensors subjected to various physical inputs like strain, displacement, velocity, acceleration, force, pressure and torque. It explains the mechanics of the measurement of these physical quantities. A detailed discussion is also made on the fabrication of these sensors. In some cases basic instrumentation based on mechanical application is explained. All these basic aspects will give a clear idea of an interested student to manufacture any instrument. Some related problems are solved to explain more clearly the outcome of observations for getting a clear idea of the concern topics on instrumentation. Lastly measurement uncertainty is discussed to express any observational error with some confidence level.

PRE-REQUISITES

Mathematics: General Mathematics (Class XII)
Physics: Engineering Physics (First year, B.Tech)
Engineering Mechanics (First year, B. Tech)

UNIT OUTCOMES

List of outcomes of this unit is as follows:

U2-O1: Predict the response of sensors to various inputs
U2-O2: Construct a conceptual arrangement of necessary instruments
U2-O3: Realise the output of the instruments and the way to express them
U2-O4: Describe the order and methodology for sensor installation
U2-O5: Differentiate between types of sensors and their modes of operation and measurement

Unit-2 Outcomes	**EXPECTED MAPPING WITH COURSE OUTCOMES** *(1- Weak Correlation; 2- Medium correlation; 3- Strong Correlation)*							
	CO-1	**CO-2**	**CO-3**	**CO-4**	**CO-5**	**CO-6**	**CO-7**	**CO-8**
U2-O1	1	1	1	1	1	3	3	-
U2-O2	1	1	1	1	1	2	2	1
U2-O3	2	1	2	1	1	1	1	-
U2-O4	-	-	-	-	-	-	-	2
U2-O5	3	-	1	1	-	1	1	1

2.1 RESPONSE OF SENSORS

The definition of sensors according to the Instrument Society of America is "a device which provides a usable output in response to a specified measurand. Here the output is an 'electrical quantity' and measurand is a physical quantity, property or condition which is measurand. Typically, a sensor transforms a desired parameter into a measurable electrical signal. The input parameter may be mechanical like motion displacement, speed, velocity, force, acceleration, torque etc. There are some terminologies that are discussed below.

2.1.1 Terminology

Following terminology is used to assess the measurement of a signal.

Range and span: Range is the lowest and highest values of the signal. Span is defined as the difference between the highest and lowest values of the signal. Input full scale is the span of the input signal, whereas the output full scale is the span of the output signal. Dynamic range is the ratio of the largest to the lowest value of the signal. An example is shown below to explain the terms clearly.

Ex. A displacement sensor is designed for an input of -50 mm to 50 mm leading to an output 2.5 V to 1.0 V. Determine range, span, input full scale, output full scale and dynamic range of this sensor.

Ans:
Given, input displacement values : -50 mm to 50 mm
Output voltage : 2.5 V to 1.0 V
Range = - 50 mm to 50 mm
Span = 50 – (-50) = 100 mm
Input full scale = 100 mm
Output full scale = 2.5 – 1.0 = 1.5 V
Dynamic range = 2.5 / 1.0 = 2.5.

Example 2.1

Error and Accuracy: Error is the difference between the measured value and the true value of a parameter. Percentage error is obtained as the ratio of the error to the span expressed as a percentage. Accuracy is the closeness of the measured value to the true value. An example is shown below to explain the terms clearly.

Ex. A displacement sensor is designed for an input of -50 mm to 50 mm leading to an output 2.5 V to 1.0 V. Because of errors, the accuracy in sensing is ±2.0 mm. Obtain the percentage error and accuracy in output.

Ans:

Given, Accuracy in sensing = ±2.0 mm

Percentage error in input = 2/(-50+50)*100 = 2%

Accuracy in output = ±0.03V

Percentage error in output = 0.03/(2.5-1)*100 = 2%

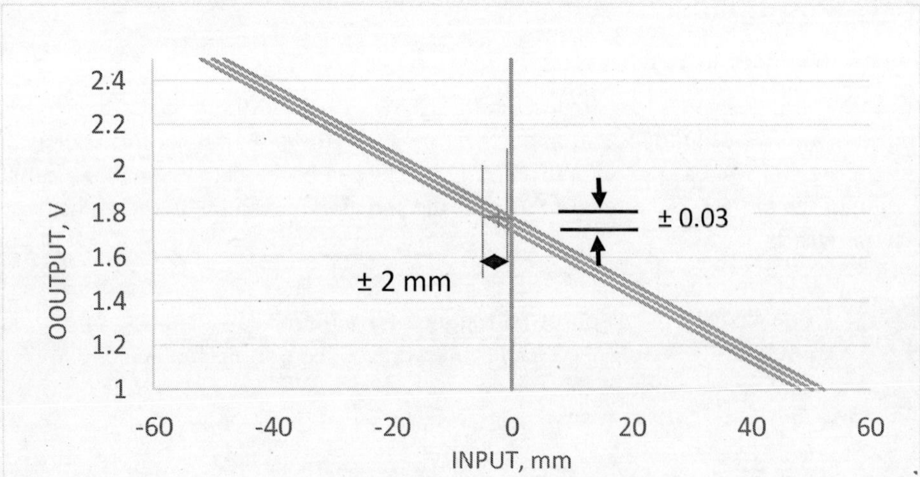

Example 2.2

Fig. 2.1: Error and accuracy of output

Hysteresis : Hysteresis is a phenomenon when a sensor does not show the same output corresponding to the same input while increasing or decreasing. Hysteresis occurs due to loose connections, magnetic fields, heat generation, etc. Fig 2.2 shows that the sensor reading is not same for a same input. The difference is called the phenomenon of hysteresis.

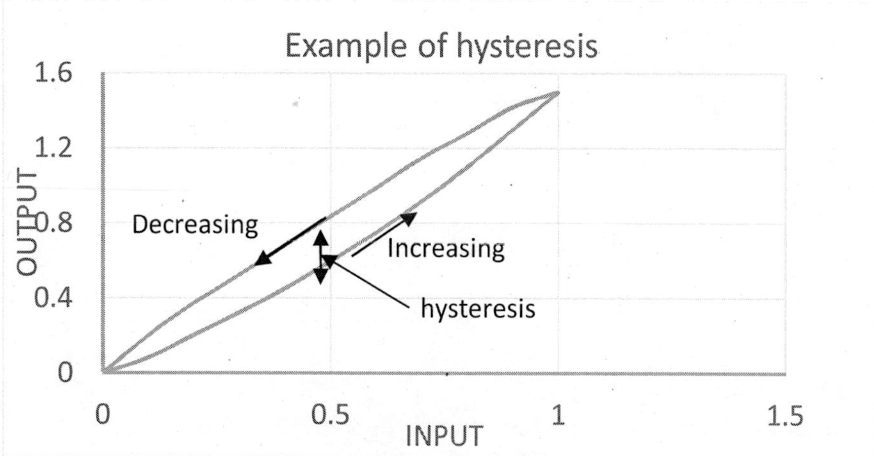

Fig. 2.2: Example of hysteresis

Nonlinearity : when the output does not vary linearly with the input data, a nonlinearity in the output results. There are two ways by which the non-linearity can be expressed. One way is the measure of deviation of the input-output relationship from the linear best fit curve of the data points. The deviation is normally measured at the mid-point of the range. In Fig 2.3 this deviation is shown as L1. The other way is the measure of deviation between the tangent at the midpoint of the range and the straight line joining the extreme points of the range. This deviation is shown as L2 in Fig 2.3

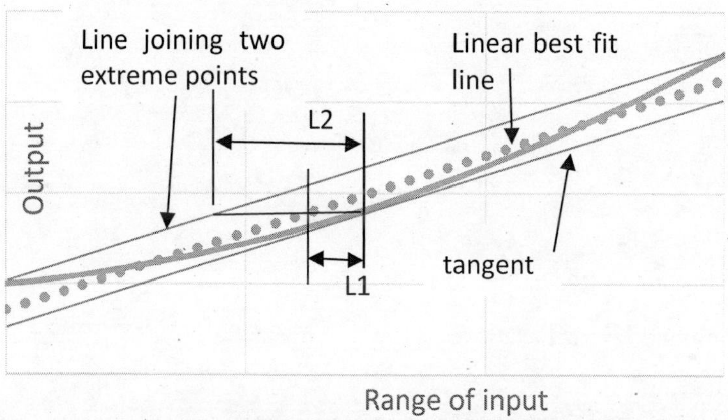

Fig. 2.3: Non-linearity of the input-output results

Ex. A displacement sensor is designed for an input range of 0 to 50 mm. Following relationship between input and output is obtained during calibration. Does the relationship is linear ? If not what is the deviation from the linear relationship and what is the maximum deviation?

Input, mm	0	5	10	15	20	25	30	35	40	45	50
Output, V	0	0.45	1.30	2.55	4.2	6.25	8.7	11.5	14.8	18.5	22.5

Ans:

The relationship is shown in the following figure. The relationship is not linear. The Linear best fit curve is shown as dotted line.

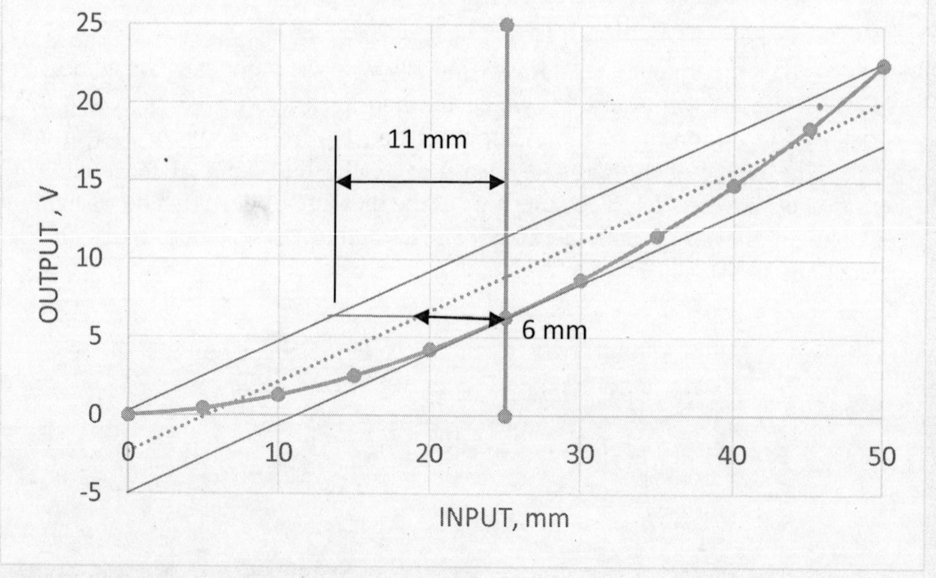

Deviation from the linear best fit = L1 = 6 mm

Total deviation = L2 = 11 mm

Example 2.3

Fig. 2.4: Example of deviation and total deviation

Dead band or dead zone : It is the range of input where there is no output. A sensor must not be operated in this zone. Fig 2.5 shows the dead zone in an output-input curve.

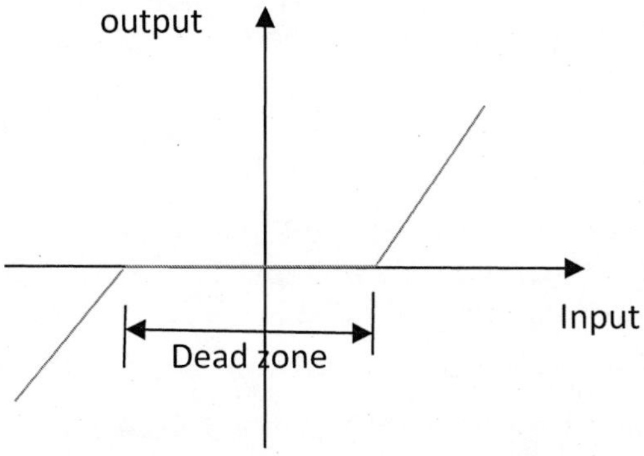

Fig. 2.5: Location of dead zone

Repeatability : It is the property of a sensor to represent the same value under identical conditions when measured at different times. Repeatability is also called reproducibility.

Reliability : This is a statistical term meaning ability of a sensor to perform its stated function without failure for a given number of cycles or a short duration under the normal working condition.

Resolution : It is the minimum increment in stimulus to which a sensor can respond. In other words, the minimum increment in input results in the least measure in output.

Frequency response : The ability of a sensor to receive the harmonic input data. It may also be defined as the plot between magnitude and frequency. For example, suppose a sensor measures change in temperature. The minimum measurable output is 0.1V. The resolution of the sensor will be the minimum change in input (i.e. temperature), which makes a 0.1 V change in the output. Fig 2.6 shows the details of the frequency response of a sensor.

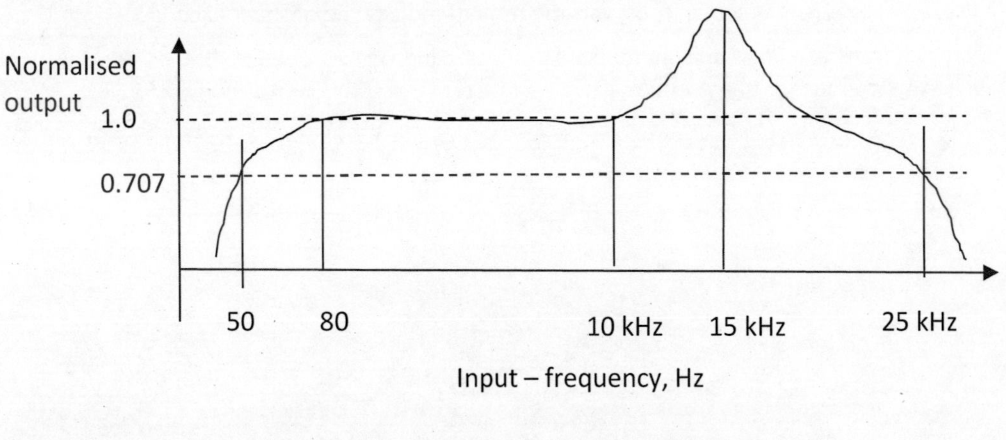

Fig. 2.6: Frequency response of a sensor

The graph shows that the maximum output occurs at a frequency of 15 kHz. This is called resonance frequency. The frequency response curve shows two frequencies 50 Hz and 25 kHz where the output is $1/\sqrt{2}$ i.e. 0.707 times the maximum value. These are called cutoff frequencies. The difference between two cutoff frequencies is called the frequency band. In this case the bandwidth = 25000-50 = 24950 Hz = 24.95 kHz. The response shows two frequencies 80 Hz and 10 kHz where the output is nearly constant, called flat frequency range. Thus, the flat frequency range for the present case = 10000-80 = 9920 Hz

Sensitivity : The sensitivity of a sensor is defined as the change in output with a unit change in input. The sensitivity represents the slope of the transfer function. For example, if the resistance is the output and the temperature is the input, and they are related with the following expression:

$$\mathbf{R = aT + b} \quad (2.1)$$

Then the sensitivity is obtained as change in resistance with unit change in temperature.

Sensitivity is expressed as

$$\frac{dR}{dT} = a \quad (2.2)$$

2.2 RESPONSE OF SENSORS TO VARIOUS INPUTS

There are various inputs which can be sensed by sensors. However, we shall concentrate on displacement, velocity, acceleration, force, torque, strain, density and vibration as inputs. All these measurements are widely used in the Civil Engineering field.

2.2.1 Displacement as input

Mechanical displacement is converted into an electrical variable due to a change in resistance in an electric circuit. An example of a mechanical displacement sensor is a Linear Variable Differential Transformer (LVDT). This is the most commonly used instrument for the measurement of linear displacement. A simple displacement sensor is the variable inductance type where the induction varies according to the displacement. This can be achieved either by varying the self-inductance or by varying mutual inductance of the two coils.

An LVDT is a transformer with a single primary coil P and two secondary coils S1 and S2, wound on a hollow cylinder as shown in Fig 2.7

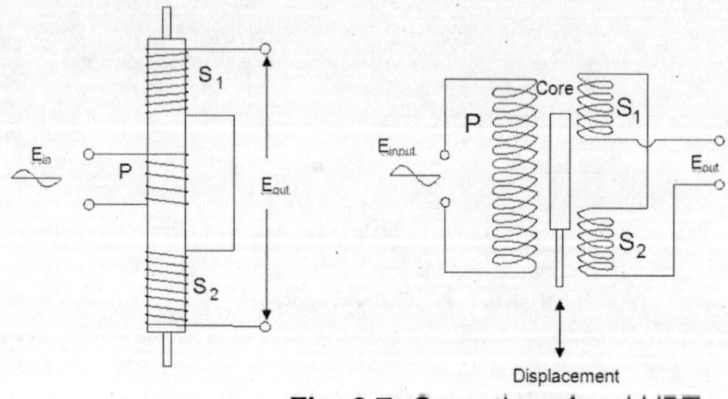

Fig. 2.7: Operation of an LVDT

The secondary coils have equal turns but are connected in series in opposite direction so that opposite emf is induced in them. The primary coil is connected to an AC supply. An iron core moves inside the hollow tube. When the iron core is in the central position, equal but opposite emf is induced in the two secondary coils and the net output is zero volt. The object of which displacement is measured is connected with this iron core. If the core is displaced more towards the left coil, more magnetic flux links with the left coil than the right coil. The emf induced in the left coil, Es1, is therefore larger than the emf induced in the right coil, Es2. The output voltage will be difference between the two secondary coil voltages and is in phase with the voltage of the left hand coil.

Similarly, when the iron core moves to the right side more magnetic flux will be linking to the right coil and the emf induced in the right coil will be more than that in the left coil. The output voltage will be the difference between the two voltages induced in the secondary coils and is in phase with the emf of the right coil. The amount of output voltage of an LVDT is a linear function of the core displacement within a limited range of motion. Thus, the movement of the core can be summarised into three cases

Case 1: In this case the flux linkage with S_1 will be more as compared to S_2. This means that emf induced in S_1 will be more than that induced in S_2.Hence $E_1 > E_2$ and the net differential voltage which is the output voltage equals to $E_0 = E_1 - E_2$ which is positive. This means that E_0 is in phase with E_1.

Case 2: The core at the null position: When the core is at the null position, the flux linkage with both the secondary windings will be the same. So the induced emf(E_1 and E_2) in both the windings will be the same. Hence net differential output voltage $E_0 = E_1 - E_2$ will be zero.

Case 3: In this case the flux linkage with S_2 will be more as compared to S_1. This means that emf induced in S_2 will be more than that induced in S_1.Hence $E_2 > E_1$ and the net differential voltage which is the output voltage equals to $E_0 = E_2 - E_1$ which is positive. This means that E_0 is in phase with E_2.

All the three cases are shown in Fig 2.8.

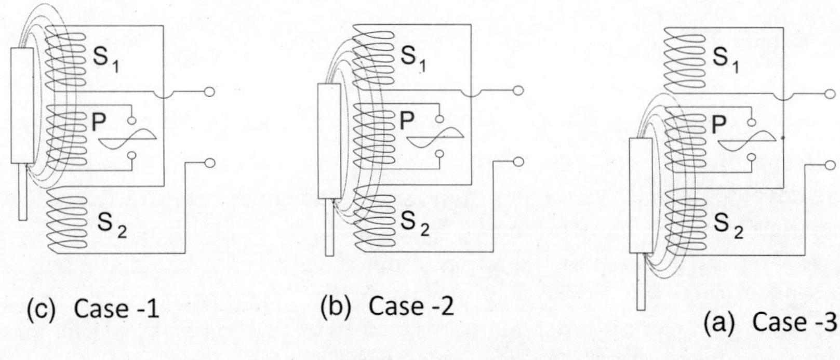

Fig. 2.8: All the three cases of an LVDT

Thus, from the above discussion it is found that the magnitude of E_0 is a function of the distance moved by the core and its polarity or phase indicates as to in which direction it has moved.

If the core is attached to an object, the magnitude of E_0 gives the position of that object.

Rotary variable differential transformer (RVDT)

It is similar to LVDT except that its core rotates about a shaft when there is a flux cut between primary and secondary coils. The operation of a RVDT is shown in Fig 2.9

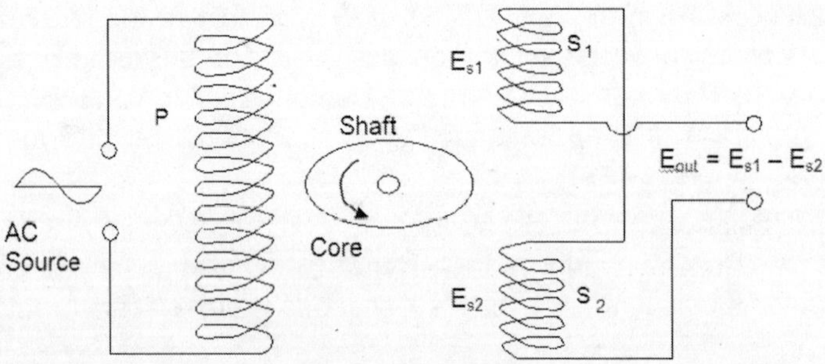

Fig. 2.9: Operation of RVDT

At the null position of the core the output voltages from the two secondary coils are equal and opposite. When the core rotates about the shaft there is a differential voltage output. The more is the angle of rotation the more is the magnitude of differential voltage. A clockwise rotation produces an increasing voltage of a secondary winding of one phase while counter clockwise rotation produces an increasing voltage of opposite phase. Hence the amount of angular displacement and its direction may be ascertained from the magnitude and phase of the output voltage of the transducer.

2.2.2 Pressure as an input

Working of a pressure sensor using a piezo-electric crystal is rather simple.J. Cuire discovered certain crystals in 1880. When a thin slice of these crystals are subjected to a mechanical force, then opposite charges are developed on the two faces of the slice. The larger is the mechanical force, the larger is the electric potential between the two faces. It is observed that if the direction of the mechanical force is reversed that is the direction of deformation is reversed, the polarity of the potential produced across the faces also gets reversed. This phenomenon is called Piezoelectric effect and the crystals are called piezoelectric materials. The examples of piezoelectric materials are Quartz, Rochelle salts and Tourmaline

The crystal is placed between a solid base and a force summing member it can be easily understood that when a pressure is applied, the force summing member deforms the crystal whereby an output voltage is generated. The range of output may be 1 to 30 mV and no external power is required. Its size is also small and the construction is simple. The piezo-electric pressure sensor consists of pairs of quartz discs placed between a flat metal face called a load plate and transducer housing. A preloading spring of stiffness k_1 is used to hold this arrangement.

When the piezoelectric crystal is placed in a thin walled tube which is enclosed in a thin diaphragm with an inbuilt cooling unit the external pressure is split into two parts. One part of the external force P is borne by crystal pile P_1 and the remaining part P_2 is borne by the thin walled tube or a diaphragm. When the crystal pile is compressed by Δx then we get

$$P = P_1 + P_2 = k_1\Delta x + k_2\Delta x = (k_1 + k_2)\,\Delta x$$

$$\frac{P}{P} = \frac{dR}{dT} = \frac{dR}{dT} = \frac{dR}{dT} \tag{2.3}$$

2.2.3 Force as an input

Load cells are the devices which convert weight or force into electrical outputs. The load cells use strain gauges for measuring the force. A pictorial view of the load cell is shown in Fig. 2.10 (a). It consists of a steel cylinder on which four identical strain gauges are installed. The gauges R_1 and R_3 are along the direction of applied load and the gauges R_2 and R_4 are along circumference of the steel cylinder. All the four strain gauges are connected electrically to the four limbs of a wheatstone bridge. The bridge circuit is shown in Fig 2.10 (b).

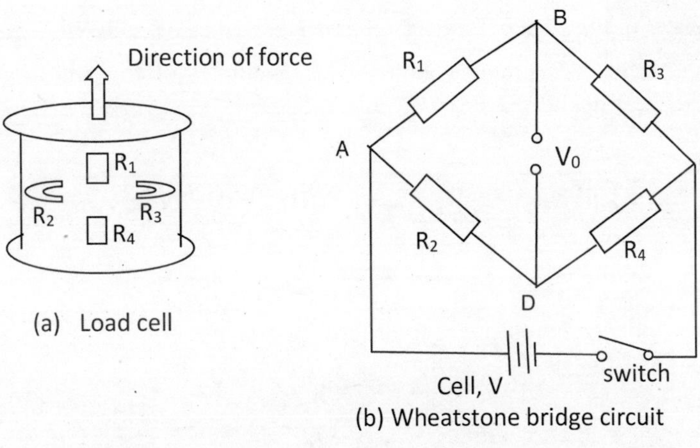

Fig. 2.10: Strain gauge load cell

When there is no load on the cell, all four gauges have the same resistance, i.e. $R_1 = R_2 = R_3 = R_4$ Obviously the terminals B and D are at the same potential, the bridge is balanced and the output voltage is given by

$$V_{AB} = V_{AD} = \frac{v}{2} \tag{2.4}$$

$$V_{AB} - V_{AD} = V_0 = 0 \qquad compression$$

$$R_2 = R_3 = R + dR \qquad tension$$

Potential at terminal B, $V_{AB} = \frac{R_1}{R_1+R_3} V = \frac{R-dR}{(R-dR)+(R+\mu\, dR)} V = \frac{R-dR}{2R-(1-\mu)dR} V$

Potential at terminal D, $V_{AD} = \frac{R_2}{R_2+R_4} V = \frac{R+\mu\, dR}{(R+\mu\, dR)+(R-dR)} V = \frac{R+\mu dR}{2R-(1-\mu)dR} V$

The new output voltage, $V_0 + dV_0 = \frac{R-dR}{2\,R-(1-\mu)dR} V - \frac{R+\mu\, dR}{2R-(1-\mu)dR} V$

$$= \frac{2(1+\mu)dR}{R}\frac{V}{4}$$

If the cell starts from an unloaded condition, i.e., $V_0 = 0$, the change in output voltage due to the applied load is given by

$$dV_0 = \frac{2(1+\mu)dR}{R}\frac{V}{4} \tag{2.5}$$

Thus, voltage is a measure of the applied load.

Ex. Obtain the output voltage due to application of a force 1 kN to a strain gauge load cell arranged with four identical strain gauges with the following particulars :

Diameter of the steel cylinder = 65 mm, Resistance of each gauge = 100 ohm;

Gauge factor = 2; supply voltage = 8 V; Modulus of elasticity of steel = 210 GPa; Poisson's ratio = 0.3

Ans:

Stress = Load/area = $\frac{1*10^3}{\frac{\pi}{4}*(0.065)^2} = 301.51 * 10^3$ N/m²

Strain = $\frac{stress}{Modulus\ of\ elasticity} = \frac{301.51*10^3}{210*10^9} = 1.436 * 10^{-6}$

Ratio of change in resistance to original resistance

$= \frac{dR}{R}$ = gauge factor * strain

$= 2.0 * 1.436 * 10^{-6} = 2.872 * 10^{-6}$

Output voltage = $dV_0 = \frac{2(1+\mu)dR}{R}\frac{V}{4} = 2 * (1+0.3) * 2.872 * 10^{-6} * \frac{8}{4}$

$= 14.93 * 10^{-6}\ V = 14.93\ \mu V$

Example 2.4

2.2.4 Torque as an input

The torque or twisting moment is calculated when a force is applied tangentially to a shaft. The torque is calculated by multiplying the force by the radius of the shaft. The torque measuring sensor may be constructed with strain gauges placed in a Wheatstone bridge principle.

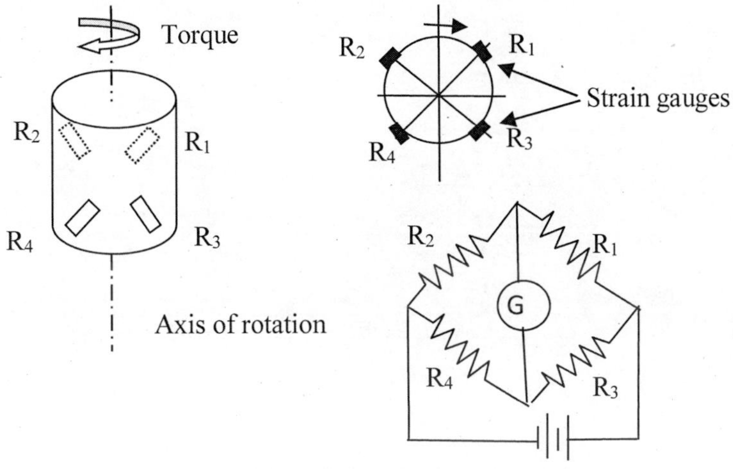

Fig. 2.11: Torque sensor

Figure 2.11 shows the installation of strain gauges around a shaft for measuring a torque. The strain gauges are mounted on a 45^0 helix with the axis of rotation. Due to torsion the diagonal with strain gauges R_2 and R_3 are in tension and the diagonal with R_1 and R_4 are in compression. For a mechanical torsion meter, the following relationship holds good:

$$\frac{T}{J} = \frac{\tau}{r} = \frac{C\theta}{l} \qquad (2.6)$$

Where T = Torque in the shaft

J = polar moment of inertia of the shaft

τ = maximum shear stress induced in the shaft

r = radius of the shaft

C =Modulus of rigidity of the shaft material

θ = Angle of twist

l = length of the shaft

from above, the torque can be measured provided it is calibrated with the angle of twist through following formula

$$T = \frac{CJ}{l}\theta = constant * \theta \qquad (2.7)$$

Ex. A torque meter transmits a power of 50 kW due to running at a constant speed of 1200 rpm. The meter uses two pairs of strain gauges, each having a resistance of 80 ohm and a gauge factor of 2. The induced voltage in the Wheatstone bridge circuit is 10V and the maximum strain in the gauges is 0.001. The shear modulus of the shaft material is 120 GPa. Calculate :

(a) The diameter of the shaft
(b) The output voltage.

Ans:

If T is the torque, the power can be expressed as

$$P = \frac{2\pi NT}{60} \text{ in watts}$$

$$60 \text{ x } 10^3 = \frac{2\,\pi * 1200 * T}{60}$$

$$T = 477.7 \quad \text{N-m}$$

$$\text{Shear modulus} = \frac{shear\ stress,\ \tau}{shear\ strain}$$

$$\text{Shear stress }, 120 * 10^9 * 0.001 = 120 * 10^6\ \ N/m^2$$

From $\frac{T}{J} = \frac{\tau}{r}$ we get $\frac{477.7}{\frac{\pi}{32} * d^4} = \frac{120 * 10^6}{\frac{d}{2}}$ i.e. d = 0.027 m = 27 mm

To obtain output voltage :

$$\frac{dR}{R} = change\ in\ resistance\ over\ original\ resistance$$

$$= gauge\ factor * strain = 2 * 0.001 = 0.002$$

$$\text{Output voltage} = 2 * \left(\frac{dR}{R} * \frac{v_0}{4}\right) = 2 * 0.002 * \frac{10}{4} = 10 * 10^{-3} = 10\ mV$$

Example 2.5

2.2.5 Velocity as an input

The velocity can be measured either with a moving magnet or coil.

In the moving magnet type one permanent magnet moves within a fixed coil. The magnet is connected to the device whose velocity is to be measured. The arrangement is shown in Fig 2.12. When the magnet moves, a voltage is induced in the coil. The velocity is directly proportional to the induced voltage.

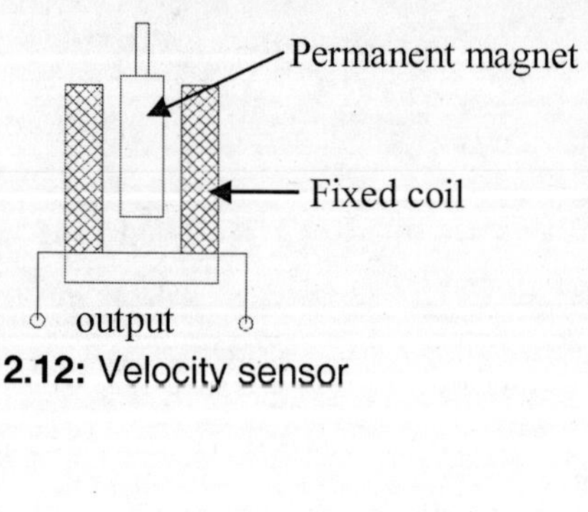

Fig. 2.12: Velocity sensor

In the moving coil type velocity sensor, the coil moves in the magnetic field of a permanent magnet. The voltage induced in the coil gives a direct measure to the velocity of movement.

2.2.6 Acceleration as an input

The acceleration of a vibrating system can be measured with a piezoelectric material. The piezoelectrical material has a property of generation of voltage due to deformation and vice versa. The working of the sensor is discussed below.

One piezoelectric crystal is sandwiched between two electrodes. One mass is placed over an electrode. The unit is fastened to a base which is to be placed on the moving body. When the mass exerts a pressure on the crystal, the crystal is deformed and a voltage is generated. Now, suppose the sensor is placed over a vibrating surface which is subjected to acceleration. When the surface moves downwards, there is a relaxation of the stress on the piezoelectric crystal. This relaxation of the crystal induces a change in voltage. The opposite deformation will be induced to the crystal when the surface moves upward. It is known that the force = mass x acceleration. Since the mass is constant, hence the decrease or increase in force is proportional to the acceleration. The output voltage is correlated with the acceleration. Fig 2.13 shows the working of an accelerometer.

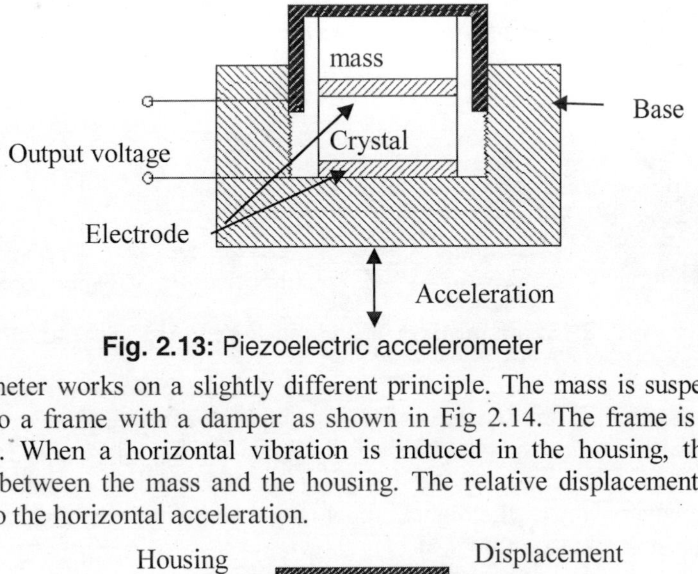

Fig. 2.13: Piezoelectric accelerometer

A seismic accelerometer works on a slightly different principle. The mass is suspended by a spring and is fixed to a frame with a damper as shown in Fig 2.14. The frame is connected to the vibrating unit. When a horizontal vibration is induced in the housing, there is a relative displacement between the mass and the housing. The relative displacement is sensed and is proportional to the horizontal acceleration.

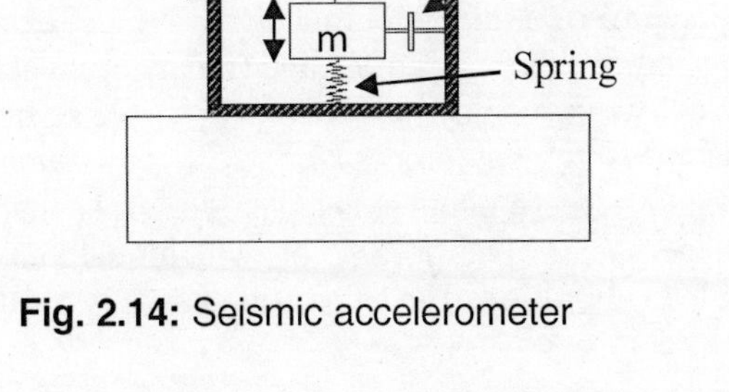

Fig. 2.14: Seismic accelerometer

2.3 CONCEPTUAL INSTRUMENTATION AND MONITORING PROGRAM

In order to have an effective measure of a physical quantity, the concept of a relevant instrument is very important. An instrument is a device that transforms a measurand into a suitable form of measurement. For an example, if measurand is length then the measurement is the number of units (millimetre, inches, etc) that represent the length. Fig 2.15 shows a pictorial representation of the concept of an instrument. The left side shows a physical process which is to be measured. This physical process is called a measurand. The measurand has some observable physical variable X which is fed to a sensor as an input. The sensor converts the input into a signal variable output.

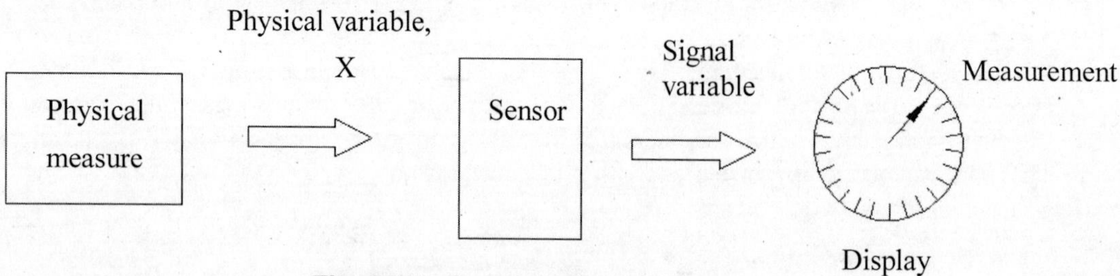

Fig. 2.15: Concept of an instrument

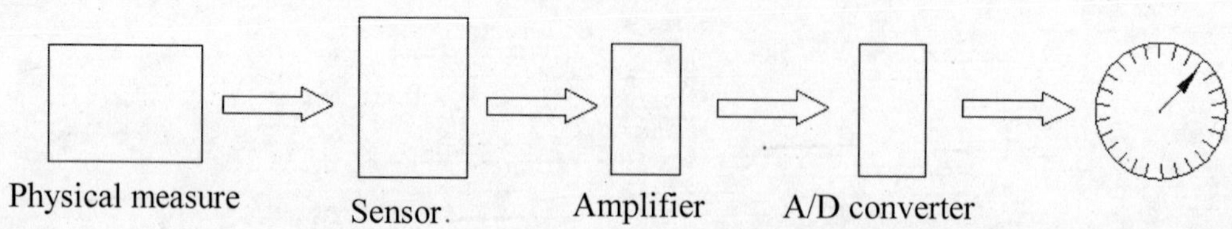

Fig. 2.16: Installation of an amplifier for a small output

In case the signal output from the sensor is small, an amplifier is required to amplify the output. The output can be sent to a display device or can be stored for future use. In case the amplified output is not digital rather analog, the output may be converted to a digital one by installing one analog to digital converter. The digital output may then be sent to a computer or may be stored or may be transmitted to another device as shown in Fig 2.16. A sensor may be an active sensor or a passive sensor. An active sensor directly sends a measuring signal and receives back the signal which is then converted into an output signal by the sensor itself. For example, a radar or a sonar is an active sensor where a radio wave (radar) or an acoustic wave (sonar) is sent to an object and subsequently received by the sensor. A passive sensor on the other hand does not add any energy to the measurement process rather it may remove the input energy during its operation. For example, a thermocouple is a passive sensor that converts a temperature measure to an electric voltage signal. The process by which an input signal is converted to a digital output signal is called calibration. The calibration curve is obtained by providing a known input signal and observing the output signal. The calibration curve is shown in fig 2.17 and is normally linear passing through the origin. The curve is linear and is not valid for an input signal in the nousen-linear zone. The valid limit of the input signal for the calibration curve is also shown in Fig 2.17 The calibration curve is used to obtain the output signal within the linear variation zone. Thus, all the instruments show one working range, which is up to the linear range between input and output signals. If the instrument is loaded with an input value greater than this elastic range, the sensitivity of the instrument may enter into the non-linear range and the output signals for the subsequent tests will show inconsistent results. The sensitivity of the device is defined as the slope of the calibration curve. Thus, sensitivity is the ratio of the change in output values to the change in corresponding input values upto the elastic limit.

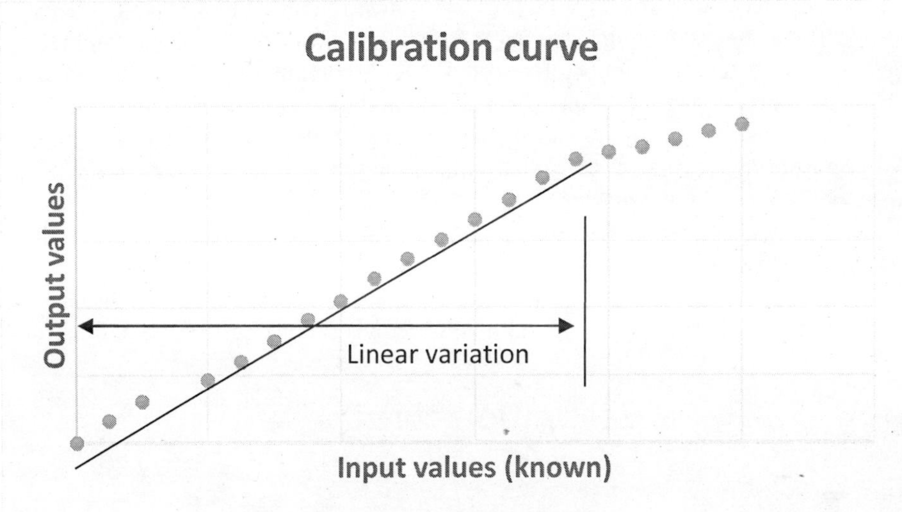

Fig. 2.17: Calibration curve

Interfering inputs:

There may be some interfering inputs which will automatically modify the output. One of the interfering objects may be temperature. The input signal is now a combination of original measurand and the additional interfering object. Within the linear range, the relationship between the input and output signals may be linear, but the sensitivity will be different. Fig. 2.18 shows the interference of an additional input signal with respect to the original signal of non-interference. Thus, many instruments have a working range of temperature. For a force measuring sensor, the vibrations may be an interfering object.

Accuracy and error:

Accuracy means how closeness is the measured value to the actual value. Error is an estimate of the difference of the true value from the measured value of the measurand. The true value is obtained either from an absolute value or from an agreed upon value. One reason for inaccurate measurement is the interfering object. The error for any measurement consists of systematic error (also called bias) and random error (also called noise). There are varieties of factors which contribute to the systematic and random errors. The systematic error may be due to any of the following reasons:

a) Miscalibration: the miscalibration is mainly due to interfering inputs. For example, when the temperature is an interfering input, any measurement taken at a temperature other than that during calibration of the sensor will induce a miscalibration error.
b) Aging factor: Aging of the components also induces the systematic error. Aging changes the workability of the sensor and hence, the calibration. This systematic error may be minimised with periodic calibration of the equipment.
c) Invasiveness: A systematic error may be induced due to the performance of the equipment itself. This is termed as invasiveness. In many instruments, invasiveness is a major concern and is difficult to be removed. For an example when a large warm thermometer is used to measure the changes in temperature of a small volume of liquid, the warmness of the thermometer may change the input temperature and thus give an inaccurate measurement. For an electronic system, the energy drain may be reduced by using a high input impedance.
d) Change in physical property of signal path: Sometimes changes in the physical property of the signal path also changes the measured value. For a mechanical device, frictional resistance may change the property whereas for an electrical device, changes in the resistance, induces the systematic error.
e) Parallax error : this error is because of the accurate measurement by the observer. When the observation is not normal to the scale, the oblique vision induces a systematic error

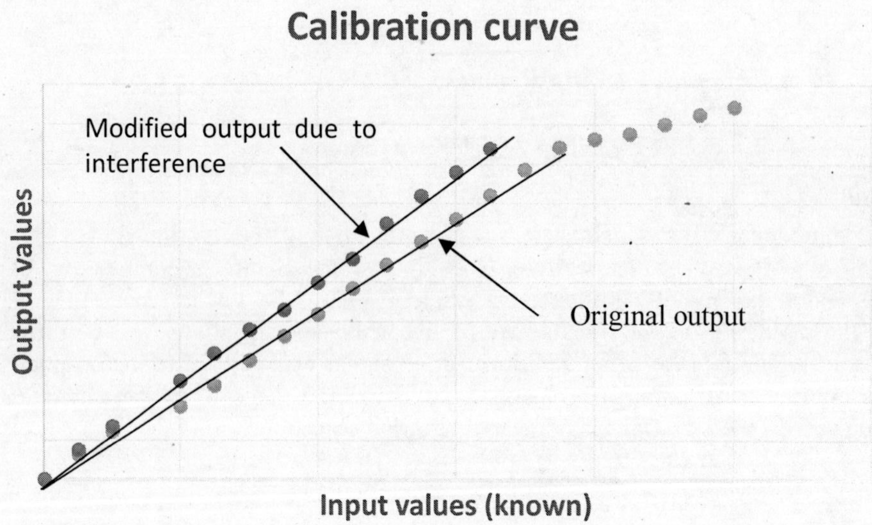

Fig. 2.18: Interference of an additional input on the calibration curve

Systematic error sources :

There are many factors which contribute to the continuous error to the output signal. For example, if heat is the modifying agent, the output signal will be modified. There will be continuous errors in the output signal. The output measurement is also modified, in case the working temperature is different from that recorded during the calibration process. The ageing is another factor.

2.3.1 Selection of instrument and its application

For any engineering job it is very essential to select appropriate instruments and also to plan the monitoring programme. Many a time it is observed that the engineer is misguided by the manufacturing firms unless he is having a very specific knowledge on the objectives of the study and the final interpretation of measured data. Thus, the following steps are to be followed for a proper selection of instrument and monitoring of the observations:

a) Define project conditions: It should contain the answer to the question "what are the engineering conditions of a project? For an example, in a geotechnical project the project conditions include subsurface stratigraphy, geotechnical properties, ground water location, conditions of nearby structures, other construction details.
b) Define the purpose of instrumentation
c) Select variables to be monitored: For any project, the variables to be measured are to be ascertained. For example, in any geotechnical project, the variables to be measured include groundwater level, pore water pressure, earth pressure, vertical and horizontal deformations, tilt, load and strain
d) Make prediction of behaviour: It is necessary to predict the outcome of any measurement. Unless the expected value of any variable is known it will be difficult to prepare the specification of the instrument. Moreover, the accuracy of the result can also be ascertained with the preliminary knowledge.
e) Devise solution to problems that may arise during observations : while observing the data , some problems may arise which are to be solved with previous experience or expertise
f) Select instrument : this is very important to select a proper instrument. The instrument should be reliable, durable, environment friendly and simple in operation. The cheap instrument may not always give an accurate result. Table 2.1 gives a list of equipment to measure different variables of a geotechnical problem and their principle of operation.
g) Plan recording the factors that may influence the data : There may be various factors which effect the data, like temperature, humidity, snow, rainfall etc. These factors are to be recorded during the measurement.
h) Ensure the procedure for correctness of reading : Correctness of result is very important. Hence periodic calibration, inter laboratory comparisons and uncertainty measurements are necessary.
i) Select instrument location ; The instrument is to be properly located. Sometimes, ongoing construction may demand a change of location of a working instrument. Hence, the instrument should be able to maintain the accuracy of results in a new location. Otherwise, an additional and adjustable instrument is to be installed for the new location.

j) Plan installation: Proper installation of the instrument is very essential. The manufacturer guidelines are to be followed. If necessary the base construction is to be completed before the purchase of the instrument

Table 2.1: Selection of an instrument for a geotechnical project

Variable to be measured	Types of instrument	Measurement technique
Pore pressure	Piezometers a) Open standpipe b) Twin tube hydraulic c) Pneumatic d) Vibrating wire strain gauge e) Bonded resistant strain gauge	a) Water enters into the standpipe from a definite location, other portion is sealed (Fig. 2.19) b) A porous tip connected with two plastic standpipes, one standpipe is connected to a pressure gauge and the other is for flushing any gas bubble (Fig 2.20) c) A pneumatic piezometer contains a flexible diaphragm, water pressure from soil acts on one side and the gas pressure on the other side as shown in Fig 2.21. During measurement, compressed nitrogen gas enters through an inlet tube forces the diaphragm. If the gas pressure is more than water pressure, the diaphragm moves in and compressed gas comes out through a vent pipe. The gas supply is closed, the diaphragm returns to its original position and the pore pressure becomes equal to the gas pressure which is measured in a pressure gauge. d) A vibrating wire strain gauge piezometer consists of a prestressed wire which connects to the centre of a flexible diaphragm as shown in Fig. 2.22. A change in pore pressure changes the tension in the wire. The tension is measured by plucking the wire which vibrates in a magnetic field. The frequency of the output voltage is equal to the frequency of vibration of the wire. A calibration curve gives the pore pressure from the range of voltage frequency. e) In bonded resistant strain gauge piezometer, an electrical pressure transducer with a diaphragm is in direct contact with pore pressure. The pore pressure induces a strain on the diaphragm which is directly measured by the strain gauges (Fig 2.23).
Earth pressure	There may be two types of measurements: (a) Measurement of	In general the earth pressure cell consists of a liquid filled chamber subjected to earth pressure. The pressure in the liquid is sensed by any type of pore pressure transducer as mentioned above. For an example, in a

	pressure within a soil mass (b) Measurement against the face of a structural element. Commercially available earth pressure cells include pneumatic, hydraulic, vibrating wire strain gage, bonded resistance strain gage, unbounded resistance strain gage	pneumatic earth pressure cell the liquid filled chamber is connected to the twin tube pneumatic pressure unit as shown in Fig.2.24. Most manufacturers provide calibration charts prepared during loading with air or water, but in general, the calibration curves are to be periodically checked.

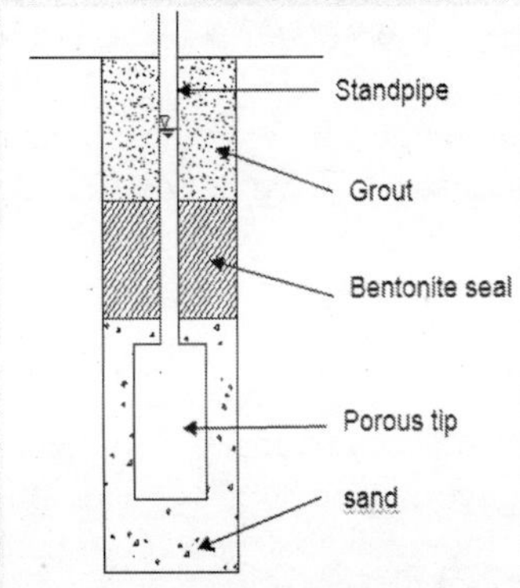

Fig. 2.19: Open stand pipe piezometer

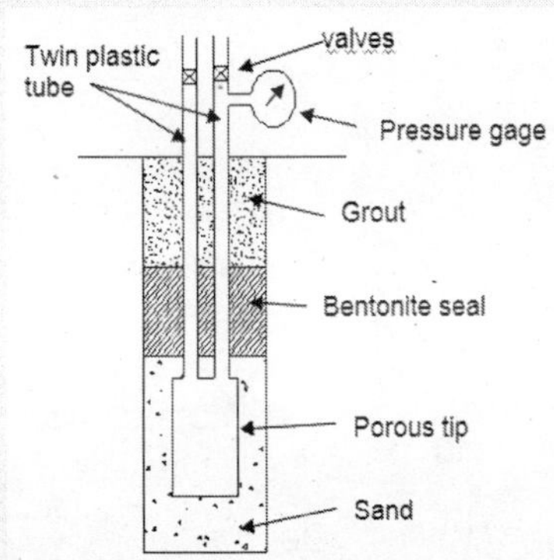

Fig. 2.20: Twin tube hydraulic piezometer

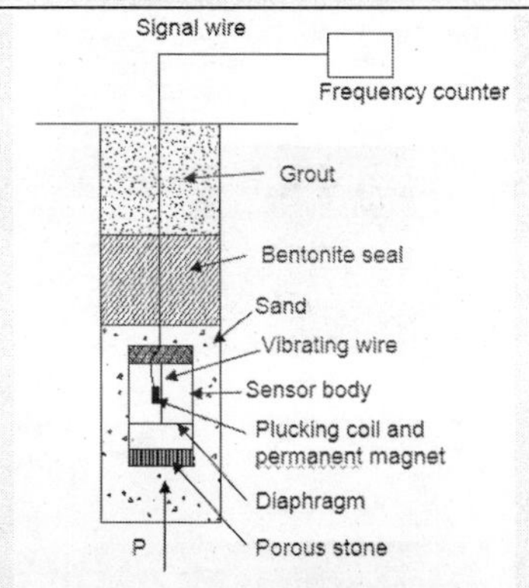

Fig. 2.22: Vibrating wire strain gage piezometer

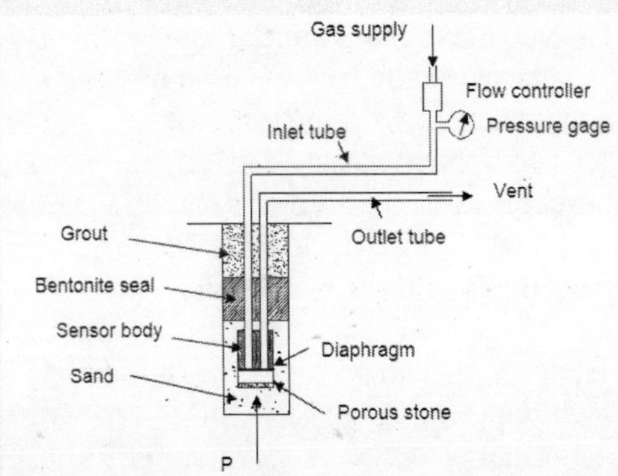

Fig. 2.21: Pneumatic piezometer

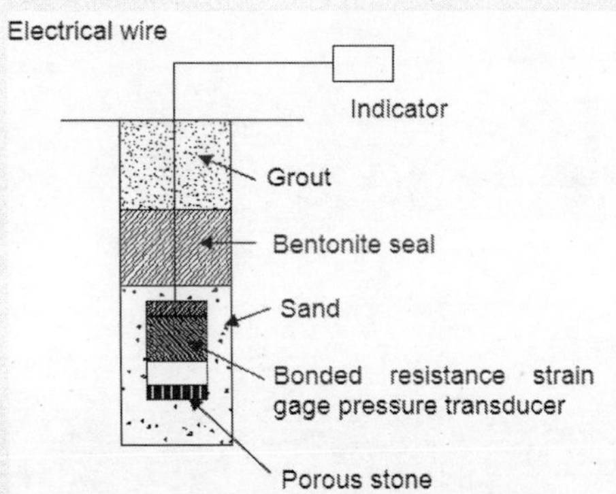

Fig. 2.23: Bonded resistance strain gage piezometer

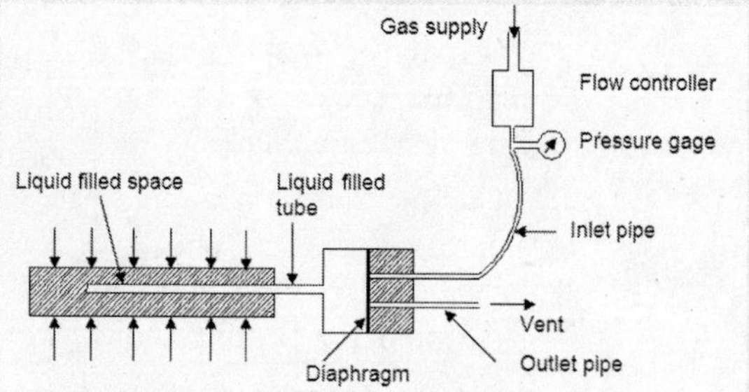

Fig. 2.24: Pneumatic earth pressure cell

Deformation	There are a large number of instruments for measuring deformation like a) Survey instruments b) Portable deformation gage c) Single point monuments d) Vertical pipe settlement gage e) Remote settlement gage f) Heave gage g) Inclinometer h) Extensiometer i) Soil strain gage	a) Survey instruments: They include total station, theodolite, dumpy level, photogrammetry etc b) Portable deformation gage: they are used for measuring cracks in buildings or on the ground surface c) Single point monuments: they are used for surface settlement measurement. Two types are common settlement rods and control stakes. d) Vertical pipe settlement gage : The settlement gage consists of a square plate of steel on which a steel riser pipe is attached. In some cases, steel prongs are used instead of steel plates. One spiral foot gage is also used in place of square plate when it is intended to measure settlement in soft soil and at variable depths as shown in Fig 2.25. e) Remote settlement gage: this type of gage is usually a manometer which can be operated from a distance. (Fig 2.26) f) Heave gage: The main purpose of this type of gage is to measure the heaving of soil at the bottom of an open cut. The working principle is explained in Fig. 2.27. A conical steel rod is installed in a slurry filled borehole at the bottom of excavation. A probing rod of known length is lowered down the borehole is placed above the conical point. The elevation of the top of the road is determined by optical levelling. The heaving can be easily measured. g) Inclinometer: It consists of a pipe with internal longitudinal guide grooves which is installed in a vertical borehole. A probe containing an electrical tilt sensor is lowered down the pipe on the end of a graduated electrical cable as shown in Fig 2.28. The electrical cable is connected to a remote readout

device indicating the tilt of the torpedo with respect to the vertical. Tilt readings and depth measurements enable the alignment of the grooved pipe to be determined. Changes in the alignment provide horizontal movement data.

h) Extensiometer: They are used to measure movements in rock and soil. They are installed inside a borehole and measure axial dimensional change of the borehole. They consist of anchors with attached rods or wires. The rods or wires extend to a head at the mouth of the borehole, where any movement of the anchors is sensed by measuring the corresponding movement of the rod or wire. A typical single anchor extensiometer is shown in Fig.2.29.

i) Soil strain gages: It is used for measuring strain within an earthfill. The device uses the principle of electrical inductance between two coils embedded within the soil. The two coils may be related to each other in orthogonal, coaxial or coplanar configuration as shown in Fig.2.30. One coil is connected to an oscillator to produce an electromagnetic field in the surrounding soil. An electric current is induced in the mating coil, the magnitude is a function of spacing between the two coils. Soil deformation is obtained from the change in spacing between the coils after installation. These devices can measure static and dynamic strains and are not affected by soil composition, water content, temperature etc.

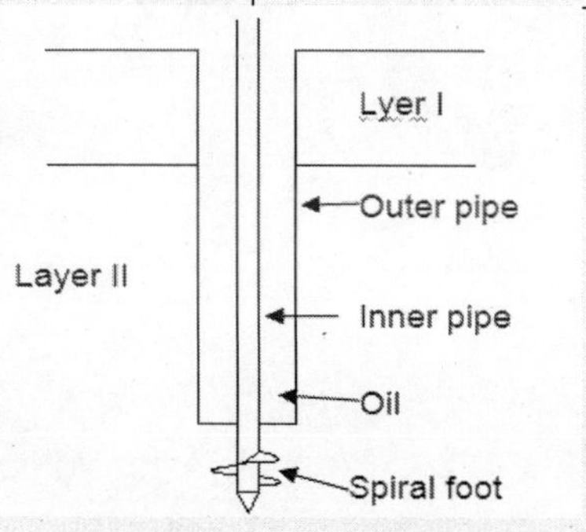

Fig. 2.25: Spiral foot gage

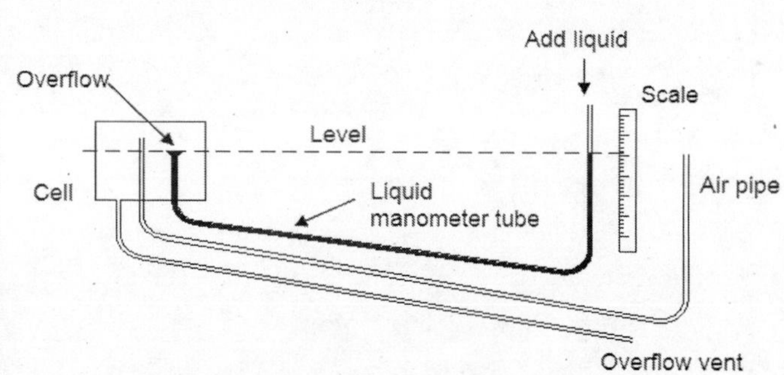

Fig. 2.26: Remote settlement gage

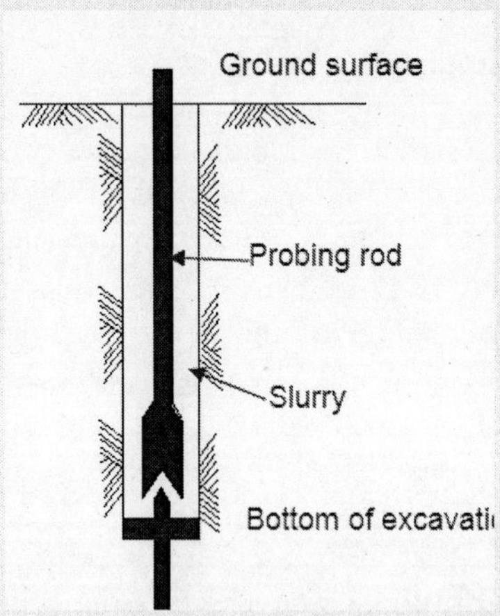

Fig. 2.27: Mechanical gage

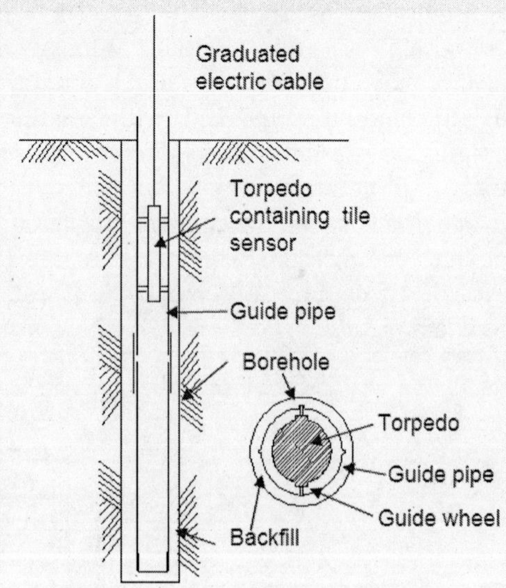

Fig. 2.28: Inclinometer

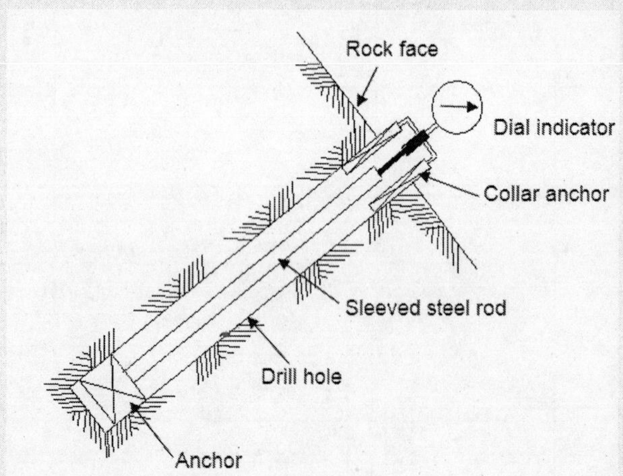

Fig. 2.29: Borehole extensiometer

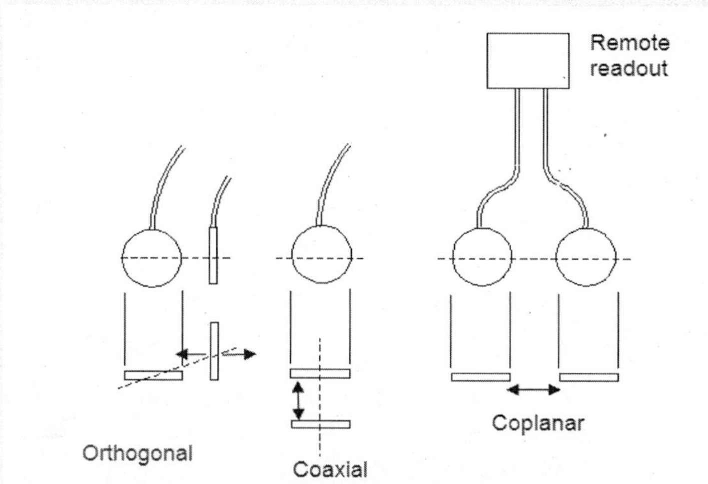

Fig. 2.30: Soil strain gage

Load and strains in structures	Five types of load cells can be used a) Tell tale b) Mechanical c) Hydraulic d) Vibrating wire strain gage e) Electrical resistance strain gage	a) Tell-tale: This type of load cell consists of an unstressed steel rod installed within a sleeve attached to the structure. One end of the rod is attached to the structure and the deformation of the other end is measured with respect to a reference point on the structure. The change in distance with respect to the original length of the rod gives the strain. Load is then calculated from the elastic modulus of the structural material and the cross section. b) Mechanical load cell consists of an elastic cup spring that is deformed during the load application. Deformation is sensed by a dial indicator and calibrated to a load. c) Hydraulic load cells consist of a flat oil chamber trapped between two bearing plates. A rock bolt may be attached at the centre. The pressure generated in the chamber can be measured either directly with a pressure gage or remotely by electrical or pneumatic means. d) Vibrating wire strain gage uses three or more vibrating wire strain gages to measure the deformation of the load bearing member. Ouput

		from each vibrating wire strain gage may be measured and averaged. e) Electrical resistance strain gage load cell: It consists of a steel or aluminium cylinder. The loads act on the ends of the cylinder, the compression is measured by the electrical resistance strain gages attached to the cylinder or bonded inside the cylinder. Several strain gages are used at regular intervals around the periphery. Half the strain gages are used to measure the tangential strains and half the strain gages are used to measure the axial strains.
	Different types of strain gages are used for the structure. They include mechanical, electrical resistance and vibrating wire strain gages.	Mechanical strain gages are used for measuring small changes in length between two studs attached to the structural member. Electrical resistance strain gages are either of the bonded or welded type. Bonded strain gages are bonded to the surface of the structure using an epoxy adhesive. Vibrating wire strain gages consist of a length of steel wire tensioned between two end blocks as shown in Fig.2.31. The blocks are attached to the surface under study. The strain of the surface causes the blocks to move relative to each other and the tension in the steel wire to change. The tension is measured by plucking the wire, using an electrical coil and measuring the frequency of its vibration. The frequency change is calibrated to strain.

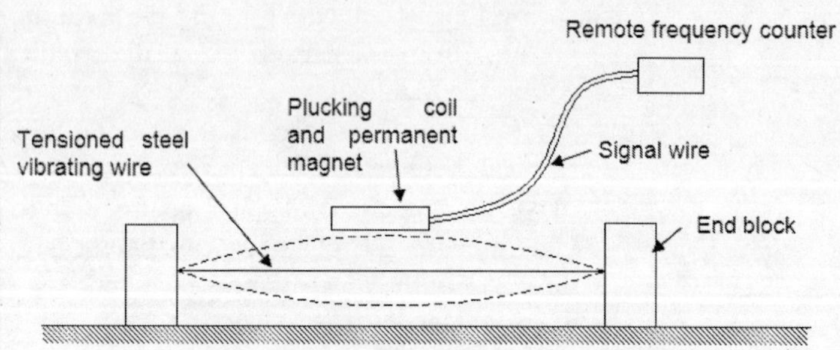

Fig. 2.31: Vibrating wire strain gage

2.4 METHODOLOGY FOR SENSOR INSTALLATION

There are common requirements for sensor installations.

1. Sensors must not protrude in any direction.
2. Sensors must not be suspended from the horizontal
3. They are to be protected from vandalism and against damage
4. Prior to installation, the ambient temperature, ambient humidity, exposure to vibration, possibility of external injury are to be assessed.
5. If the sensor is to be installed inside a false ceiling, its location is to be marked on the ceiling.
6. If the sensor is to be concealed in the wall, a plate is to be mounted near the sensor mentioning the sensor name, reference number, etc.
7. For an immersion sensor, the required length of the sensor is to be fully inserted into the water.
8. The sensors must be installed opposite to the direction of flow.
9. The sensors must not be installed normal to the direction of flow, rather to some appropriate angle.
10. Maintain a minimum clearance between the sensor and the nearby obstruction, so that the sensor can be removed easily whenever required.
11. Sensors must be installed at a safe distance (10d to 15d, d is the diameter of the pipe) from the mixing point of fluids at different temperatures.
12. Wind sensors are to be installed on the facades and are to be exposed to the main wind direction. They are not to be installed under eaves and must be assessed easily for inspection.
13. Solar sensors are to be installed on the facades directly exposed to sunlight, are to be accessible, are not to be installed in the shade of adjacent buildings
14. The room sensors are to be placed at least 1.5 m height and 0.5 m away from the adjacent wall. Room sensors must not be installed on the outer wall and directly exposed to sunlight.
15. The sensors must not be installed behind a curtain, near to a door, on a wall containing concealed hot water pipe, near a radiator, near a table lamp, on a chimney wall

Installation of some civil engineering sensors:

1. Piezometer
2. Inclinometer

Installation of piezometer

Piezometers are normally installed in a borehole. Following are the installation steps:

1. Drill a borehole and wash it thoroughly to remove all cuttings
2. Establish initial zero reading of the piezometer
3. Lower the sensor to the required level
4. Backfill the borehole with saturated clean sand around the sensor as a pore pressure collection zone
5. Bentonite is poured upto the depth of the next upper piezometer location.
6. Next the piezometer is lowered and saturated clean sand is poured around the piezometer.
7. Fill the bentonite in between the collection zones.

The inclinometer is normally installed in a borehole. Following are the steps of this installation:

1. Drill a borehole
2. Install a casing inside the borehole
3. Connect the sensor, wheel and the connector
4. Lower the assembly to the required depth.
5. Continue to add tubing, sensor and wheel assembly until the last sensors is installed
6. Record zero readings after stabilizing.
7. Working of a sensor is increased if it is properly installed. Normally the sensor is installed on a stainless steel band. Different types of band are available depending upon the background material on which the sensor is to be installed.

2.5 TYPES OF SENSORS AND THEIR MODES OF OPERATION AND MEASUREMENT

Based on the operational principle, the sensors are classified as follows:

i) Piezoresistivity sensor
ii) Piezoelectricity sensor
iii) Variable capacitance sensor
iv) Optical sensor
v) Resonant sensor
vi) Electrostatic sensor
vii) Thermal sensor
viii) Magnetic sensor

2.5.1 Piezoresistivity sensor

This type of sensors is based on the principle that some metals like copper and iron exhibit increase in resistance while in tension. The phenomenon was first observed by Lord Kelvin in 1856. He observed that the iron wires showed a higher increase in resistance than copper wires. The first application came in 1930. Instead of using the wires directly, a metal foil with the wire is made, which is then glued to a surface as shown in Fig 2.32(a). The arrangement is also called a strain gauge. The sensitivity of a strain gauge is a dimensionless term and is defined as a ratio of relative change in resistance to the strain

$$gauge\ factor\ (GF) = \frac{relative\ change\ in\ resistance}{strain} = \frac{\Delta R/R}{\Delta l/L} = \frac{\Delta R/R}{\epsilon} \tag{2.8}$$

It is known that when a tensile strain occurs, there is a decrease in lateral stain also. It can be shown that when the original length is increased, there is a decrease in width and thickness. Thus, there is also a lateral strain. The ratio of lateral strain to longitudinal strain is called Poisson's ratio. Fig 2.32(b) shows the changes in width and thickness due to change in length.

The resistance is defined as $R = \frac{\rho l}{A} = \frac{\rho l}{b\,t}$

Where, ρ is the resistivity, l =length of the element, b = width of the element and t = thickness of the element.

Differentiating : $dR = \frac{l}{b\,t}\,d\rho + \frac{\rho}{b\,t}\,dl - \frac{\rho l}{b^2\,t}\,db - \frac{\rho l}{bt^2}\,dt$

$$\frac{dR}{R} = \frac{d\rho}{\rho} + \frac{dl}{l} - \frac{db}{b} - \frac{dt}{t}$$

$$\frac{dR}{R} = \frac{d\rho}{\rho} + \epsilon_l + \gamma\,\epsilon_l + \gamma\,\epsilon_l = \frac{d\rho}{\rho} + \epsilon_l(1 + 2\,\gamma) \quad (2.9)$$

Where, ϵ_l = longitudinal strain

γ = Poisson's ratio

$$gauge\ factor\ (GF) = \frac{\Delta R/R}{\epsilon} = \frac{\frac{d\rho}{\rho}}{\epsilon_l} + (1 + 2\,\gamma) \quad (2.10)$$

Thus the gauge factor depends on three terms:

First term = $\frac{\frac{d\rho}{\rho}}{\epsilon_l}$ = resistance change due to piezo resistive effect

Second term = 1 = resistance change due to change of length

Third term = 2 γ = resistance change due to change in area

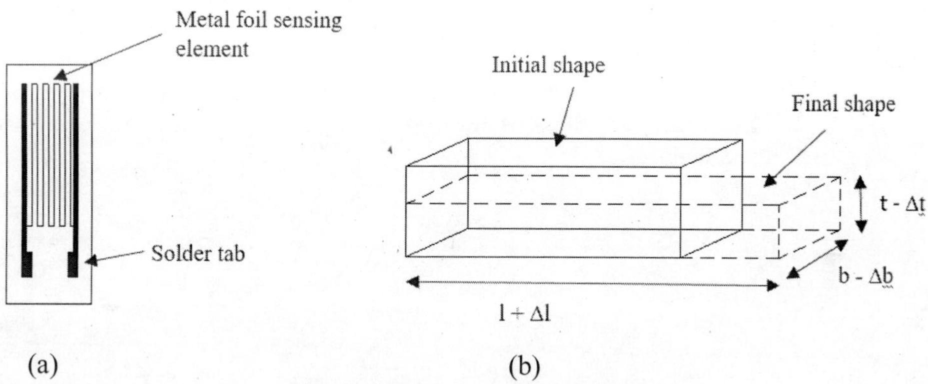

Fig. 2.32: Strain gauge sensor

The strain is usually expressed as microstrain, one micro strain = 1 μm/m

If the piezoresistive effect i.e., change in resistivity due to strain is neglected, the gauge factor is given by:

$$gauge\ factor\ (GF) = (1 + 2\,\gamma) \quad (2.11)$$

The Poisson's ratio for most of the materials lie between 0 and 0.5. Thus the gauge factor is approximately 2.0. The value of the gauge factor varies from material to material, but it is generally assumed to be constant over a range of strain. Its value is to be determined in the laboratory. Knowing the gauge factor, the strain in the member can be directly found by the change in resistance.

Properties of gage materials:

The gauge materials should possess the following desirable properties for proper functioning:

1. High resistivity
2. High mechanical strength
3. High electrical stability
4. Low temperature sensitivity
5. Low hysteresis
6. Good corrosion resistance
7. Good weldability

Proper mounting of the metal foil:

The metal foil is to be properly mounted on the surface. Following steps are to be taken:

1. The surface is to be properly cleaned with a dry cloth
2. Traces of grease or oil are to be removed using a solvent like acetone.
3. The back of the strain gauge is to be cleaned by cotton dipped in acetone to ensure that the back is totally free from grease or oil
4. Apply a proper quantity of adhesive to the cleaned surface and put the gauge over the surface. It is to be ensured that there is no air bubble between the gauge and the surface.
5. Allow the gauge to sit over the surface for eight to ten hours. If required, a small weight may be placed over a rubber sheet which in turn is placed over the gauge.
6. After the adhesive is fully cured, check the continuity of the wire by an ohmmeter and weld the electric leads.

Example 2.6

Ex. Obtain the gauge factor of a metallic rod which when stretched from 4 cm to 4.2 cm, the resistivity increases from 10 ohm-cm to 11 ohm-cm. Take Poisson's ratio of the metal = 0.25.

Ans

Longitudinal strain $\epsilon_l = \frac{0.2}{4} = 0.05$

$$\gamma = 0.25$$

$$gauge\ factor\ (GF) = \frac{\Delta R/R}{\epsilon} = \frac{\frac{1}{10}}{0.05} + (1 + 2 * 0.25)$$

$$= 3.5$$

2.5.2 Piezoelectric sensor

Certain classes of crystals exhibit generation of an electric charge, when subjected to an applied mechanical force. In the reverse effect, they deform in response to an external electric field. Quartz is one of the crystals which exhibit the piezoelectric effect. The behaviour of the quartz crystal was first discovered by Jacques and Pierre Curie in 1880. This effect is due to charge asymmetry within the crystal structure. Some examples of piezoelectric materials are:

- Rochelle salt (naturally occurring)
- Barium titanate (ceramic material)
- Lead zirconatetitanate, PZT (ceramic material)

- Polyvinylidene fluoride, PVDF (polymer material)

Figure 2.33 indicates the piezoelectric effect.

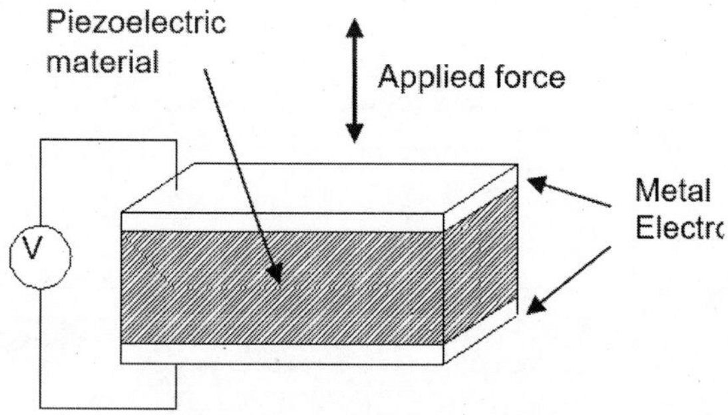

Fig. 2.33: Piezoelectric effect

There is a limiting temperature of every piezoelectric material called Curie point and the piezoelectric material does not work at a temperature higher than the Curie point. For example, the Curie point of PZT is 195^0.

Due to anisotropic nature of the charge distribution, the direction of force application is very important. There are three axes of the piezomaterial. If a force is applied along axis-3, the direction of polarization will occur across the thickness. The electrodes are to be placed across the thickness to measure the voltage generated. The piezoelectric parameter which is used to measure the voltage is called charge coefficient, d_{ij}. It is related to the charge generated on a surface across i-th axis due to a force applied along j-th axis. For the arrangement shown in fig-2.33 the charge coefficient will be d_{33}. If a force F is applied to the piezoelectric sample, the charge generated is given by

$$Q = d_{33}\, F \tag{2.12}$$

If the piezoelectric material has surface area A, thickness t and relative permittivity = ε_r, the voltage produced across the thickness is given by

$$V = \frac{Q}{C} = \frac{d_{33}\, F\, t}{\varepsilon_r \varepsilon_0 A} \tag{2.13}$$

Ex. Obtain the voltage developed across a piezoelectric sheet of dimension 10 mm x 10mm x 2mm thick with a relative permittivity = 4000. A normal force of 50 N is applied on the surface. The charge coefficient = 500 pC/N

Ans

Voltage developed, $V = \frac{Q}{C} = \frac{d_{33}\,F\,t}{\varepsilon_r \varepsilon_0 A}$

$$= \frac{500 * 10^{-12} * 50 * 2 * 10^{-3}}{4000 * 8.854 * 10^{-12} * 10 * 10 * 10^{-6}}$$

= 14.1 V

Example 2.7

For reverse action another piezoelectric constant namely voltage coefficient is used

$$\text{Voltage coefficient} = g_{ij} = \frac{d_{ij}}{\varepsilon_r \varepsilon_0} \tag{2.14}$$

In order to obtain the strain or mechanical stress following definitions are used :

$$charge\ coefficient = d = \frac{strain\ developed}{applied\ electric\ field}\left(\frac{m}{V}\right) = \frac{charge\ density}{applied\ mechanical\ stress}\left(\frac{C}{N}\right) \tag{2.15}$$

$$voltage\ coefficient = g = \frac{open\ circuit\ electric\ field}{applied\ mechanical\ stress}\left(\frac{V-m}{N}\right) = \frac{strain\ developed}{applied\ charge\ density}\left(\frac{m}{C}\right) \tag{2.16}$$

Some properties of different piezoelectric materials are shown in table 2.2(Beeby et al, 2004)

Table 2.2 Properties of some piezoelectric materials

Material	Type	d_{33} (pC/N)	Relative permittivity
Quartz	Single crystal	2	4
PVDF	Polymer	20	12
Barium titanate	Ceramic	190	2000
PZT	ceramic	300-600	400-3000

2.5.3 Capacitive sensor

The working of a capacitive sensor is rather simple. The principle lies in the fact that out of two electrodes, one is stationary and the other is movable. Fig 2.34 shows the principle of operation. It may be that the two electrodes may be stationary and the intermittent material may be moved for change in flux. For the two electrodes the capacitance is given by

$$C = \frac{\varepsilon_0 \varepsilon_r A}{d} \text{ (F)} \tag{2.17}$$

Where, ε_0 = permittivity of free space

ε_r = relative permittivity of the material between the plates

A = area of overlap between the electrodes

d = separation between the electrodes.

It can be seen that the capacitance can be changed in three ways;

(i) Changing the separation distance, d as shown in Fig 2.18(a)
(ii) Changing the overlapping area, A as shown in Fig.2.18(b)
(iii) Changing the relative permittivity of the intermittent material as shown in Fig 2.18(c)

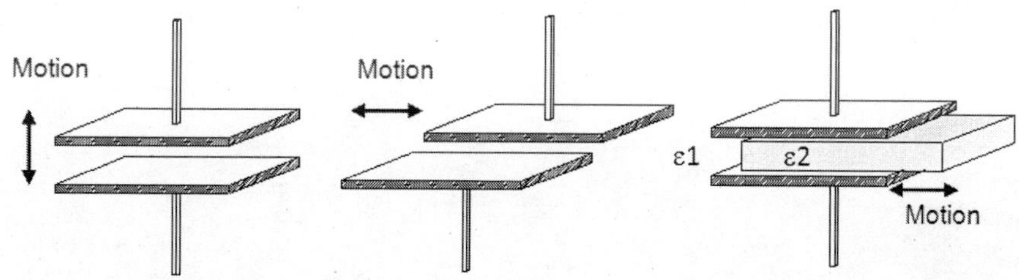

(a) Moving plate (b) Variable area (c) Moving dielectric

Fig. 2.34: Capacitor sensor

When, the spacing between two electrodes i.e., d varies, the capacitance varies non-linearity. When the overlapping area varies, the capacitance varies linearly. However, the capacitance depends upon the temperature since all three parameters like, d, A and ε_r vary with the temperature. In order to minimise the effect of the temperature, relative capacitance with three electrodes is used, as shown in Fig 2.35. In this type of capacitor, the middle electrode moves and the other two outer electrodes are fixed. When the middle electrode moves through a distance x, the differential voltage ($V_2 - V_1$) is given by:

$$(V_2 - V_1) = V_r \left(\frac{x}{d}\right) \tag{2.18}$$

Where, V_r is the supply voltage. The equation is valid for $|x| < d$

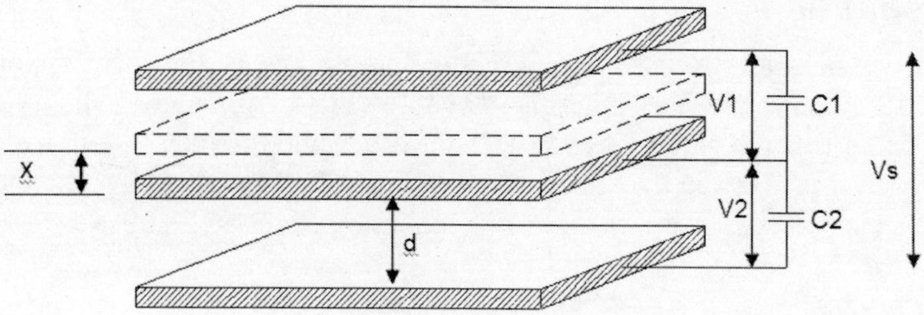

Fig. 2.35: Differential Capacitor sensor

This membrane type capacitor is used in pressure sensors and microphones.

2.5.4 Optical sensor

The optical sensors primarily work on modulating the properties of electromagnetic waves. The sensor uses the optical interface. The microsensor changes some properties of the electromagnetic wave to indicate the measurand. Followings are some of the properties of an electromagnetic wave which are altered by the optical sensors:

a) Intensity
b) Phase
c) Wavelength
d) Frequency
e) Polarization
f) Spatial position.

a) Intensity: Some optical detectors like photodiode, phototransistor, etc can detect the variations of optical intensity. A microsensor is arranged to vary the intensity of an optical signal and this variation is observed using an optical detector. A simple arrangement for the operation of an optical sensor modulating the intensity is shown in Fig 2.36. The optical source may be a Light emitting diode (LED), a laser or an incandescent lamp. The microsensor moves in respond to a measurand and this movement blocks the intensity of the light source on a photo detector. A qualitative measurement of the intensity modulation depends upon the optical beam size. The minimum beam size may be of the order of the wavelength of the optical source, and thus the movement of the microsensor may be controlled to achieve 100% intensity modulation.

b) Phase: A photodetector cannot directly respond to the variations of phase, hence it is necessary to convert the phase difference into an intensity variation before optical incidence on a photodiode. This is achieved by using one interferometer to combine one or more optical beams that are interacted by a microsensor with one or more optical beams unaffected by the microsensor. Due to interaction, the optical path length of the interacting optical beam is changed, hence the phase of the optical beam also varies. The microsensor moves in response to the measurand and reflects the optical beam thereby changes the path length. Therefore, a coherent source like a laser diode is the optical source

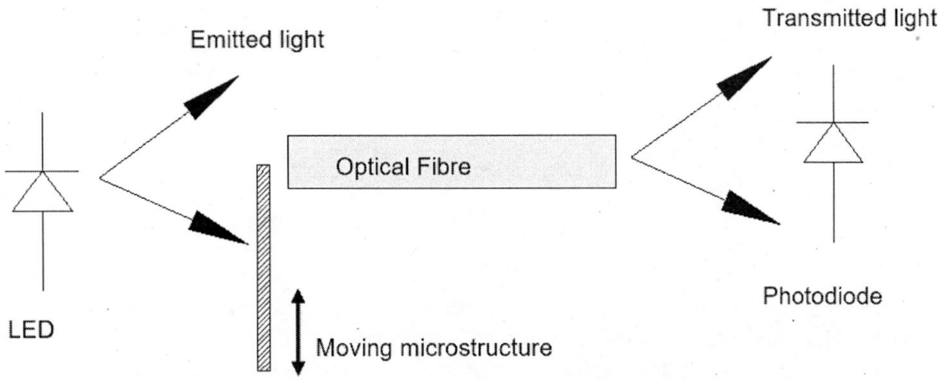

Fig. 2.36: Intensity modulated sensor

c) Wavelength: Here the optical source uses a broad spectrum. The incident light splits into different spectral segments upon interaction with the microsensor. The photodetector measures the intensity of the spectral segments. From prior knowledge of the potential modulation mechanism of the microsensor, the type of measurement and its magnitude can be obtained. A good example of the wavelength-based sensor is one based on the gas absorption, which is highly wavelength specific according to the quantity of gas present.

d) Spatial: A simple arrangement of a variation of the spatial position due to the movement of the microsensor is shown in Fig 2.37. It is clearly observed that the movement of the microsensor changes the position of the incident ray on the optical detector.

e) Frequency : when an optical radiation of frequency f is incident on a moving object with a velocity v, the frequency of the reflected radiation is changed to a frequency f_1 given by:

$$f_1 = \frac{f}{1-\frac{v}{c}} \approx f\left[1 + \frac{v}{c}\right] \tag{2.19}$$

This frequency variation is converted to intensity variation by using on interferometer which combines frequency shifted beam with non-shifted beam.

f) Polarization : Polarization means the propagation of light waves along a plane

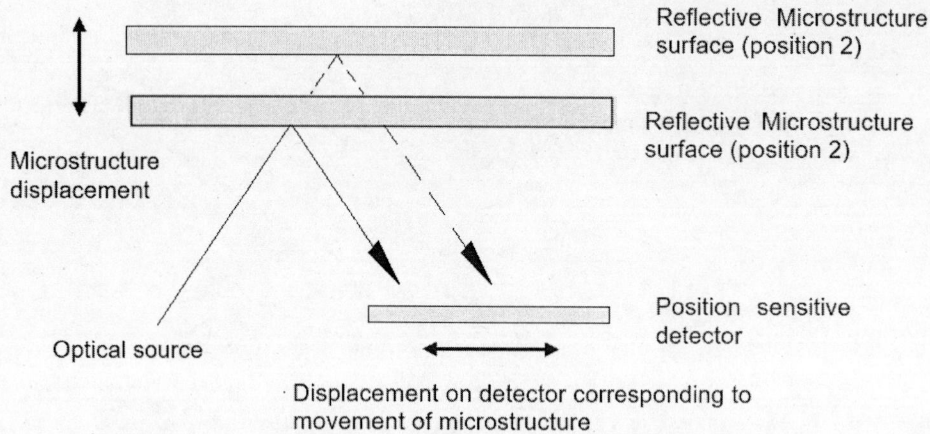

Fig. 2.37: Spatial variation in a sensor

2.5.5 Resonant sensor

A resonant sensor is a mechanical device which vibrates at a particular resonant frequency. Resonators may be fabricated from a single crystal material with micron sized materials. The resonant frequencies of these micro-resonators are extremely stable. The operation of a resonant sensor is shown with a block diagram as shown in Fig 2.38. The resonant sensor is so chosen that the natural frequency of the sensor is a function of the measurement. The measurand changes the stiffness, mass, direction of the resonator. Thus, the natural frequency of the resonator changes.

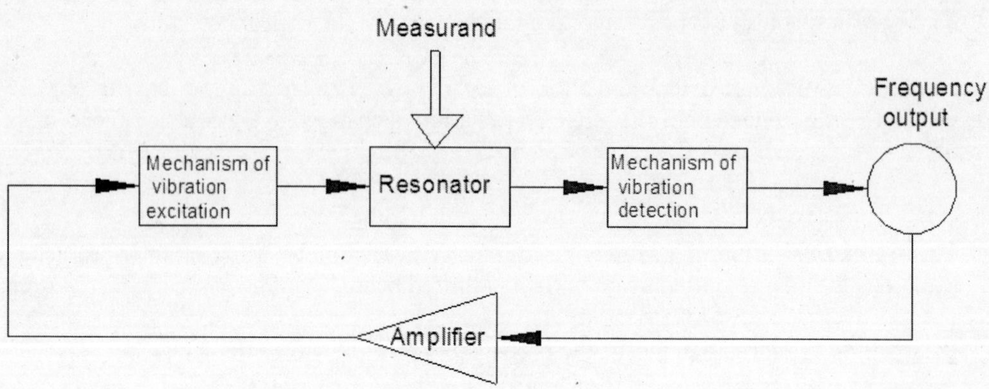

Fig. 2.38: Block diagram of a resonant sensor

2.5.6 Actuator sensor

An actuator is a device which converts electrical signals to mechanical energy. Although it is expected that the conversion should be 100%, but only 5 to 35% conversion is possible. This is because of internal and external losses. The actuator is designed with four fundamental concepts as discussed below.

a) Electrostatic : The principle of electrostatic actuator is simple. Two large plates are oppositely charged, so there will be an attractive force. This force will vary non-linearly with voltage and the spacing between the plates. A schematic diagram of the electrostatic capacitor actuator is shown in Fig 2.39.

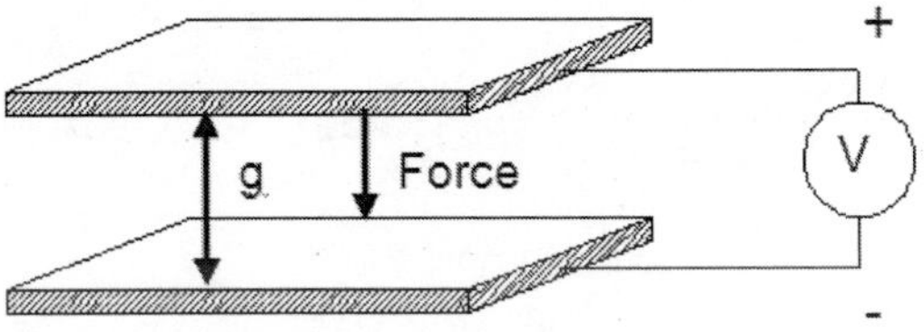

Fig. 2.39: An electrostatic capacitor actuator

If V is the voltage applied to the plates, and s is the spacing, the energy stored between the plates is given by

$$E = \frac{1}{2} C V^2 = \frac{\varepsilon_0 \varepsilon_r A V^2}{2 s} \tag{2.20}$$

The force between the plates

$$F = \frac{dE}{ds} = \frac{\varepsilon_0 \varepsilon_r A V^2}{2 s^2} \tag{2.21}$$

Another alternative to the parallel plate actuator is comb like actuator where a number of interdigitated electrodes called fingers are placed. The thickness of the fingers are small in comparison to its length and width. The attractive force is therefore due to fringing electric field instead of parallel plate field. The movement is in lateral direction. The area of overlap and the gap is fixed, so the displacement varies as the square of the voltage.

b) Piezoelectric: the working of a piezoelectric material is already discussed. When a voltage is applied across any piezoelectric electrode, the electrode will deform and the magnitude of the deformation is proportional to the voltage applied. However, the deformation is very very small, in the order of microns, even though, the applied voltage may be several hundreds of volts. The deformation is scaled down for a small piezoelectric sensor. As such, a mechanical amplifier is used to measure the deformation of a piezoelectric sensor.

c) Thermal: the thermal actuator works on the principle of coefficient of linear thermal expansion. The dissimilar materials are glued together. On heating one material expands more than the other resulting in development of thermal stresses. An example is shown in Fig 2.40. When an electric current is passed through the aluminium sheet, it gets heated up and expands more the base material. The open end moves up. This type of sensor is called a thermal bimorph.

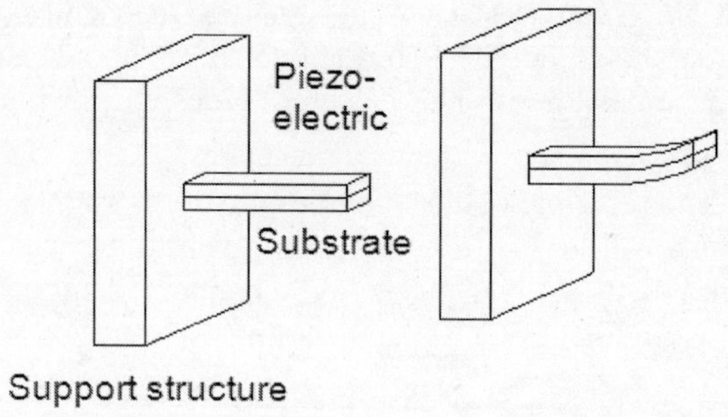

Fig. 2.40: A simple cantilever beam with a deposited piezoelectric

d) Magnetic : this type of sensors act on the principle that when a magnetic material moves in an electric field one electromagnetic force (Lorenz Force) is generated. The direction of this force is perpendicular to the current and magnetic field. The magnitude of this force is proportional to the current, length of the element and the magnetic field. Since permanent magnetic materials are not available for small scale sensors, this type of sensors are replaced with another type of sensor where the dimension of the magnetic material is changed with its magnetic state. This is called the magnetostrictive effect. Like piezoelectric material, the magnetostrictive effect is also reversible. All magnetic materials possess varying degree of magnetostrictive effect. The effect was discovered in 1847 by J.P.Joule. He observed that the length of an iron bar was changed when it was magnetised. Modern day magnetostrictive material is Terfenol-D, an alloy of terbium, dysprosium and iron. The Terfenol-D has a strain in the range of 2 x 10^{-3}. This strain is much higher than that in piezoelectric sensors. Researchers are trying to deposit thin and thick films of magnetostrictive materials onto substrates like silicon, glass and alumina. Judy et al. suggested one electroplated magnetic actuator as shown in Fig 2.41. The actuator is electroplated with 7 μm thick layer of Permalloy. The actuator is connected to a polysilicon base with a narrow beam. The actuator deflects along the direction of a magnetic field.

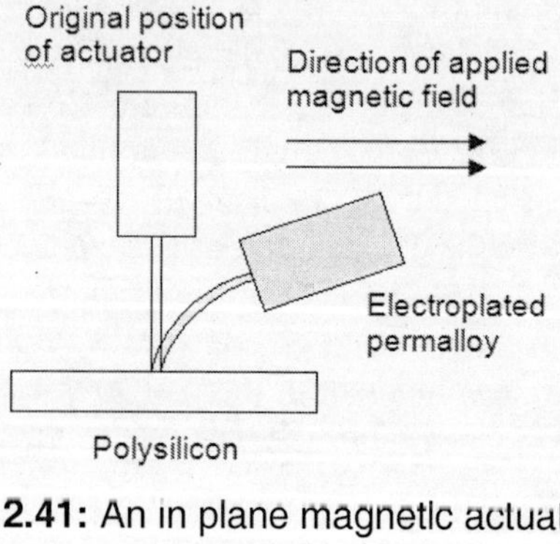

Fig. 2.41: An in plane magnetic actuator

2.6 APPROACH TO PLANNING MONITORING PROGRAMMES, DEFINE TARGETS, SENSOR SELECTION AND INSTALLATION

In the Civil engineering field, the following physical quantities are normally measured:

i) Strain
ii) Displacement
iii) Force
iv) Torque
v) Velocity
vi) Acceleration

To measure the above quantities, an appropriate type of sensor is to be selected. A small description on the selection process and working principle of the selected sensors is noted below.

a) Strain measurement
b) Displacement measurement

Strain is defined as the change in the length of a linear element divided by the original length of the element. For the measurement of strain, the usual practice is to make measurements over a short distance. This is because; it is assumed that the strain is constant over the length of a specimen. The strain is measured with a strain gauge, which gives an indication of strain at only one point. The strain gauges are normally arranged in a Wheatstone bridge circuit. There is another type of circuit used for strain gauges, it is the Ballast circuit. However, only the Wheatstone bridge circuit is discussed here

Wheatstone bridge circuit

The Wheatstone bridge technique can be used for the following two cases

a) Null mode and b) deflection mode

a) **Null mode** :

In a null mode, the galvanometer, as shown in Fig. 2.42 will show zero deflection. In this case, there will be no straining of the strain gauge R_1. The four resistances have the following relationship :

$$\frac{R_1}{R_3} = \frac{R_2}{R_4}$$

In this case, R_1 is the strain gauge resistance ($= R_g$) R_2 and R_4 are the fixed resistance and R_3 is the variable resistor.

When the gauge is strained, the resistance R_1 will change by an amount = dR_1. This disbalances the bridge equilibrium, and the galvanometer will deflect. The balance is then regained by adjusting R_3

$$\frac{R_1+dR_1}{R_3+d\,R_3} = \frac{R_2}{R_4}$$

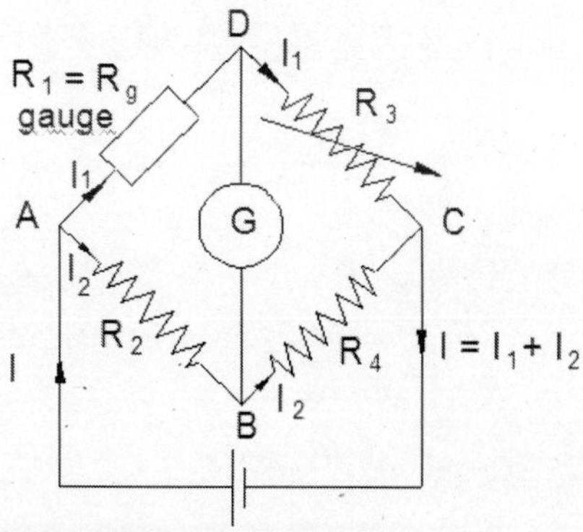

Fig. 2.42: Wheatstone bridge circuit

$$\text{Or,}\quad (R_1 + dR_1) = (R_3 + dR_3)\frac{R_2}{R_4} = (R_3)\frac{R_2}{R_4} + (dR_3)\frac{R_2}{R_4}$$

$$\text{Or,}\quad (R_1 + dR_1) = R_1 + (dR_3)\frac{R_2}{R_4}$$

$$\text{Or,}\quad dR_1 = (dR_3)\frac{R_2}{R_4}$$

If all the resistances, of a Wheatstone bridge are equal

$$R_1 = R_2 = R_3 = R_4 = R_g$$

$$\text{And } dR_1 = dR_3$$

The change in resistance dR_1 in terms of strain is given by

$dR_1 = GF \in R_g$, where, GF is the gauge factor and $\in$ is the strain

$dR_3 = GF \in R_g$, where, GF is the gauge factor and $\in$ is the strain

The above equation suggests that the change in the value of resistance is a direct measurement of strain

b) **Deflection mode**

In the deflection mode, the bridge is first balanced. The equilibrium is disturbed when the gauge is strained due to deflection. The voltage v_0 is measured in this unbalanced condition. There may be three types of bridges which are applied for deflection mode.

i) **Quarter bridge**:

In the quarter bridge, only one strain gauge is used. Other three elements of the bridge are fixed resistors. Fig. 2.43 (a) shows the position of a single strain gauge on a cantilever beam. The circuit diagram is also shown in Fig. 2.43 (b)

Ex. The following data are obtained for a Wheatstone bridge circuit :

$R_1 = 75.8\ \Omega$; $R_2 = 82.5\ \Omega$; R_3 = variable ; $R_4 = 600\ \Omega$;

a) Calculate the value of R_3 for null mode

b) Calculate the change in resistance of the gauge if the resistance is increased to 720Ω after strain is applied.

c) Calculate the strain if gauge factor is 2.2

Ans :

a) Value of R_3

We know that $\frac{R_1}{R_3} = \frac{R_2}{R_4}$

$$R_3 = R_1 \times \frac{R_4}{R_2} = \frac{75.8 * 600}{82.5} = 551.27\ \Omega$$

b) Change in gauge resistance

$Dr_3 = 720 - 551.27 = 168.73\ \Omega$

change in gauge resistance : $Dr_1 = Dr_3 \times \frac{R_2}{R_4} = 168.73 \times \frac{82.5}{600} = 23.2\ \Omega$

c) Strain

$$GF = \frac{\frac{dR_1}{R_1}}{\epsilon} \quad \text{thus strain} = \epsilon = \frac{\frac{23.2}{75.8}}{2.2} = 0.1391 = 13.91\%$$

Example 2.8

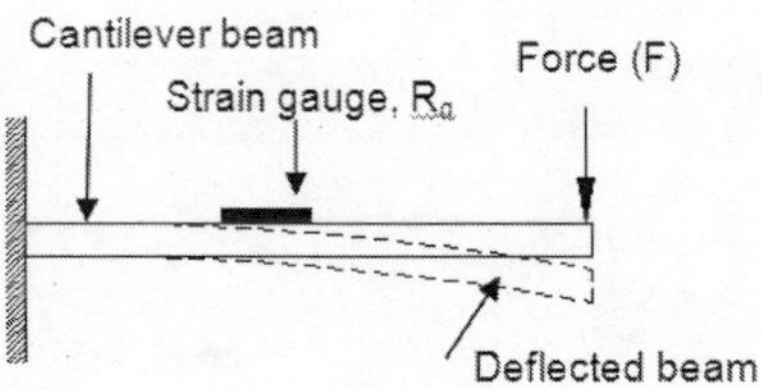

(a)

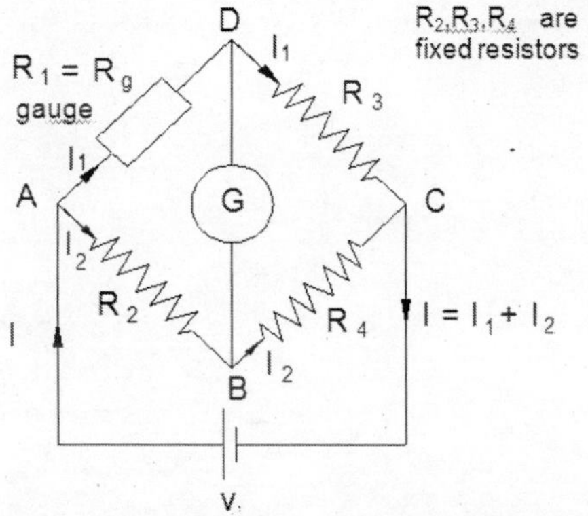

(b)

Fig. 2.43: (a) Single strain gauge on a cantilever beam; (b) Quarter bridge circuit

Assuming that the galvanometer has an infinite impedance, hence no current will flow through it.

Current along the limbs AD and DC = $I_1 = \frac{v_i}{R_g + R_3}$

Voltage drop across the gauge = $v_{AD} = I_1 R_g = R_g * \frac{v_i}{R_g + R_3}$

Similarly $I_2 = \frac{v_i}{R_2 + R_4}$

And $v_{AB} = I_2 R_2 = R_2 * \frac{v_i}{R_2 + R_4}$

Initially, consider $R_g = R_2 = R_3 = R_4 = R$

$$v_{AD} = v_{AB} = \frac{v_i}{2}$$

For such a case, voltage across B and D = $v_0 = v_{AB} - v_{AD} = 0$

Thus the bridge is balanced under unstrained condition.

Under strained condition, the gauge resistance will change to $R_g + \Delta R_g = R + \Delta R$ (say)

Voltage across the gauge = $v_{AD} = (R_g + \Delta R_g) * \frac{v_i}{R_g + \Delta R_g + R_3} = (R + \Delta R) * \frac{v_i}{(2R + \Delta R)}$

$v_{AB} = R_2 * \frac{v_i}{R_2 + R_4} = \frac{v_i}{2}$ since $R_2 = R_3 = R_4$

final output voltage = $v_0 + dv_0 = (R + \Delta R) * \frac{v_i}{(2R + \Delta R)} - \frac{v_i}{2} = v_i * \frac{\Delta R}{2\,(2R + \Delta R)}$

Assuming, $\Delta R << R$, $4R + 2\,\Delta R \cong 4R$

$$dv_0 = \frac{v_i * \Delta R}{4R} = \frac{GF}{4} \in V_i \tag{2.22}$$

Where, GF is the gauge factor. The output voltage is, therefore, proportional to strain.

Ex. A strain gauge with a resistance 120Ω and a gauge factor of 2.0 is mounted on a steel bar. It forms one arm of a symmetrical Wheatstone bridge circuit. A tensile load is applied to the bar and the galvanometer shows an output voltage 5.0 mV. If a current 15 mA flows through the gauge, calculate the mechanical strain developed in the bar

Ans:

Given, R = 120Ω ; GF = 2.0 ; $dv_0 = 5.0\ mV$; $I_1 = 15\ mA$

Voltage across the gauge = 15 x 10^{-3} x 120 = 1.8 V

Since the other resistor is also having the same resistance, hence voltage across the other resistor in the same limb as the gauge = 1.8 V

Input voltage = v_i = 3.6 V

After application of tensile force, the change in voltage = 5.0 mV

From the relationship

$$dv_0 = \frac{GF}{4} \in v_i$$

Mechanical strain $= \in = \frac{5.0 * 10^{-3} * 4}{2 * 3.6} = 2.78 \times 10^{-3}$

Example 2.9

Ex. A strain gauge with a gauge factor of 2.0 is strained to a tensile strain of 1.5 x10^{-3}. It forms one arm of a symmetrical Wheatstone bridge circuit. The value of all the resistors including the gauge is 120Ω. Calculate :

a) Change in value of variable resistance for rebalancing to null bridge circuit
b) Output voltage obtained in the galvanometer for an input voltage of 8 V.

Ans:

Given, $R = R_g = R_2 = R_3 = R_4 = 120\Omega$; GF = 2.0 ;
strain $\in = 1.5 \times 10^{-3}$, $v_i = 8V$

a) Change in the value of variable resistance, R_3

We know that the gauge factor

$$GF = \frac{\frac{dR_1}{R_1}}{\in} \quad \text{or,} \quad 2.0 = \frac{\frac{dR_1}{120}}{0.0015}$$

$$dR_1 = 0.36\ \Omega$$

We know $\frac{dR_1}{dR_3} = \frac{R_2}{R_4}$

Or, $\frac{0.36}{dR_3} = \frac{120}{120}$

$$dR_3 = 0.36\ \Omega$$

b) Output voltage in galvanometer

$$dv_0 = \frac{GF}{4} \in v_i = \frac{2}{4} * 0.0015 * 8 = 0.006 = 6\ mV$$

Example 2.10

ii) Half bridge

Here two gauges are used for strain measurement. The other two resistors in the Wheatstone bridge circuit have fixed resistances. The stain gauge – 1 is pasted on the upper side of the beam and the strain gauge -3 is pasted on the lower side of the beam. Thus the upper strain gauge is under tension and the lower strain gauge is under compression when a force is applied to the beam in the downward direction Fig. 2.28 shows the strain gauge arrangement on the beam and the half bridge circuit. In this case, the temperature effect is cancelled with R_2 = R_4 and two identical gauges on the beam.

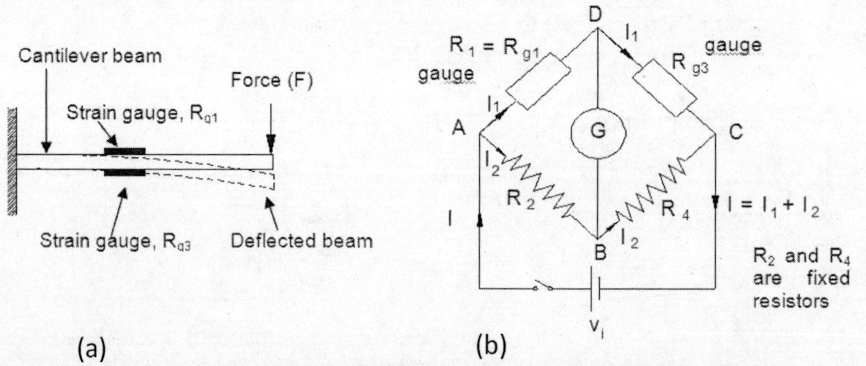

Fig. 2.44: (a) Two strain gauges on a cantilever beam; (b) Half bridge circuit

Suppose initially $R_{g1} = R_{g3} = R_2 = R_4 = R$

And under no strain condition $v_{AD} = v_{AB} = \frac{v_i}{2}$ and $v_D = v_B$ *and* $v_0 = 0$

Under the load, that is under strained condition, the gauge resistance R_{g1}increases due to tension and the gauge resistance R_{g3} decreases due to compression.

Thus resistance in gauge 1 $= R_{g1} + dR_{g1} = R + dR$

And resistance in gauge 3 $= R_{g3.} - dR_{g3} = R - dR$

$$v_{AD} = (R_{g1} + dR_{g1}) * \frac{v_i}{(R_{g1}+dR_{g1})+(R_{g3}-d\,R_{g3})} = \frac{R+dR}{(R+\Delta R)+(R-\Delta R)}v_i$$

$$= \frac{R+dR}{2R}v_i$$

$$v_{AB} = R_2 * \frac{v_i}{R_2 + R_4} = \frac{v_i}{2}$$

The change in output voltage $= dv_0 = \frac{R+dR}{2R} v_i - \frac{v_i}{2} = \frac{v_i}{2} * \frac{dR}{R}$ (2.23)

Or, $dv_0 = \frac{GF}{2} * v_i * \in$

The change in output voltage can also be expressed through changes in gauge resistances :

$$dv_0 = \frac{v_i}{2} * \frac{dR}{R} = \frac{v_i}{4} \left(\frac{dR}{R} - \left(-\frac{dR}{R} \right) \right)$$

$$= = \frac{v_i}{4} \left(\frac{dR_{g1}}{R} - \frac{dR_{g3}}{R} \right)$$

$dv_0 = \frac{v_i}{4} [(fractinal\ change\ in\ R_{g1}) - (fractional\ change\ in\ R_{g3})]$ (2.24)

iii) Full bridge

In full bridge circuit, all four resistors of a Wheatstone bridge are replaced by stain gauges. Thus, there will be two strain gauges on the upper part of the beam and two strain gauges on the lower part of the beam. Fig 2.45 shows the strain gauge arrangement and the full bridge circuit.

Initially under the unstrained condition $R_{g1} = R_{g2} = R_{g3} = R_{g4} = R$

And under no strain condition $v_{AD} = v_{AB} = \frac{v_i}{2}$ and $v_D = v_B$ *and* $v_0 = 0$

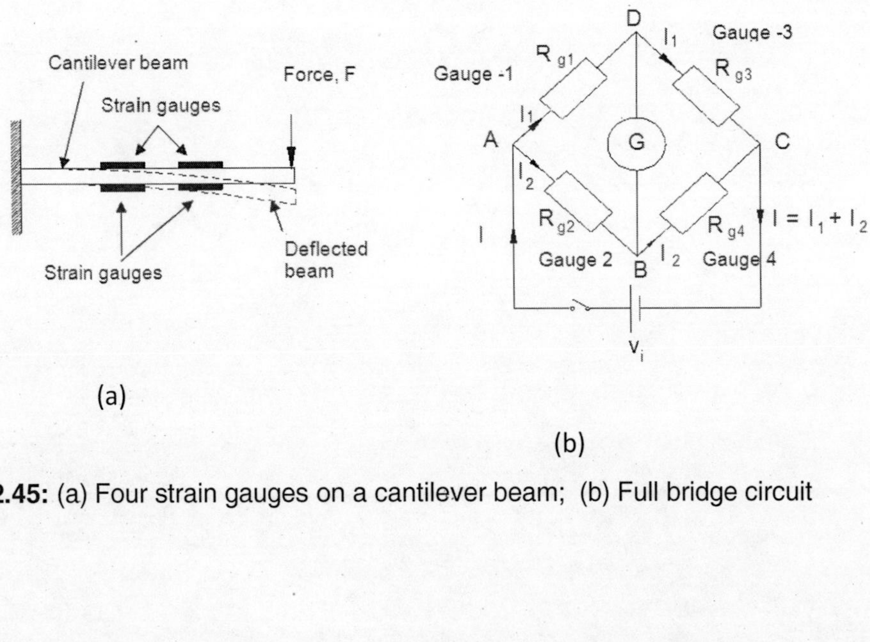

Fig. 2.45: (a) Four strain gauges on a cantilever beam; (b) Full bridge circuit

Under the load, that is under strained condition, the gauge resistances R_{g1} and R_{g4} increase due to tension and the gauge resistances R_{g2} and R_{g3} decrease due to compression.

Thus resistance in gauges 1 and 4 = $R_{g1} = R_{g4} = R + dR$

And resistance in gauges 2 and 3 = $R_{g2} = R_{g3.} = R - dR$

$$v_{AD} = \frac{R_{g1}}{R_{g1} + R_{g3}} v_i$$

$$v_{AD} = \frac{R+dR}{(R+dR)+ (R-dR)} v_i = \frac{R+dR}{2\,R} v_i \tag{2.25}$$

And

$$v_{AB} = \frac{R_{g2}}{R_{g2} + R_{g3}} v_i$$

$$v_{AB} = \frac{R-dR}{(R-dR)+ (R+dR)} v_i = \frac{R-dR}{2\,R} v_i \tag{2.26}$$

The change in output voltage can also be expressed through changes in gauge resistances:

$$dv_0 = \frac{R+dR}{2\,R} v_i - \frac{R-dR}{2\,R} v_i = v_i * \frac{dR}{R}$$

$$dv_0 = GF * v_i * \in \tag{2.27}$$

This is four times the change in output voltage due to one gauge only.

Ex. For measuring strain, four strain gauges are installed on a cantilever beam so that a full –bridge circuit of Wheatstone bridge is formed. The gauges are mounted on the fixed end of the beam. The size of the beam is 750 mm long x 70mm wide x 40 mm deep. The nominal resistance of each strain gauge is 150 Ω and the gauge factor is 2.0. The bridge supply voltage is 10 V and the measuring instrument i.e. the galvanometer has an infinitely high internal resistance. If a 10 N force is applied at the free end, calculate the sensitivity of the system in mV/N

Take, modulus of elasticity of mild steel = 210 GPa.

Ans:

Given, $R_{g1} = R_{g2} = R_{g3} = R_{g4} = 150\Omega$; GF = 2.0 ;

Length of beam = 0.75 m, width = 0.07 m, depth = 0.04 m , v_i = 10V

E= 210 GPa

Bending moment = 10 x 0.75 = 7.5 N-m

Moment of inertia = $0.07 \times (0.04)^3/12 = 3.733 \times 10^{-7}$ m^4

Let σ be the bending stress,

Therefore, $\frac{M}{I} = \frac{\sigma}{y}$

Or, $\sigma = \frac{M\,y}{I} = \frac{7.5*(\frac{0.04}{2})}{3.733*10^{-7}} = 0.40 \times 10^6$ N/m^2

Strain = $\in\ = \frac{\sigma}{E} = \frac{0.40*10^6}{210*10^9} = 1.9*10^{-6}$

Thus $\frac{dR}{R} = GF * \in\ = 2.*1.9*10^{-6} = 3.8*10^{-6}$

$dv_0 = v_i * \frac{dR}{R} = 10*3.8*10^{-6} = 38*10^{-6}$ V

$dv_0 = 38*10^{-3}$ mV

Sensitivity = $38*10^{-3}/10 = 38 \times 10^{-4}$ mV/N

Example 2.11

Displacement measurement is one of the prime measurements in the civil engineering field. Thereare many applications like building settlement, embankment settlement, laboratory tests, strain measurement, angular tilt, etc. Displacement means moving from one position to another position for a distance or angle. The displacement can be measured after a specific time that is the measurement is independent of time. This is called static measurement. In another type, the displacement can be measured over a time continuously that is the measurement is dynamic. In this chapter, static measurement is discussed. During the measurement of a displacement, the selection of an appropriate sensor is very important. The following questions are to be answered before selecting a particular type of sensor :

a) How big is the displacement and what is the type, linear or angular?
b) What are the desired accuracy and resolution?
c) How the measured object is made of (plastic, metal, fluid, ferromagnet, etc)?
d) What is the environmental condition during the measurement?
e) What is the permissible mechanical wear of the machine?
f) What is the target cost ?

An appropriate choice of the sensor will increase the efficiency of measurement.

Thus, to select an appropriate sensor, care must be taken so that the change in position of the measurand can be measured through a change in voltage. This can be done in a wire resistance. Since the resistance is directly proportional to the length, hence the movement of the measurand can be used to change the length of the wire. Due to a change in resistance, there will be a change in voltage. For this measurement a D.C. voltage is required. Referring to Fig 2.46 for the full length of the resistance, D, the potential drop is E. For any other length, d, the potential drop is given by $V = E\frac{d}{D}$

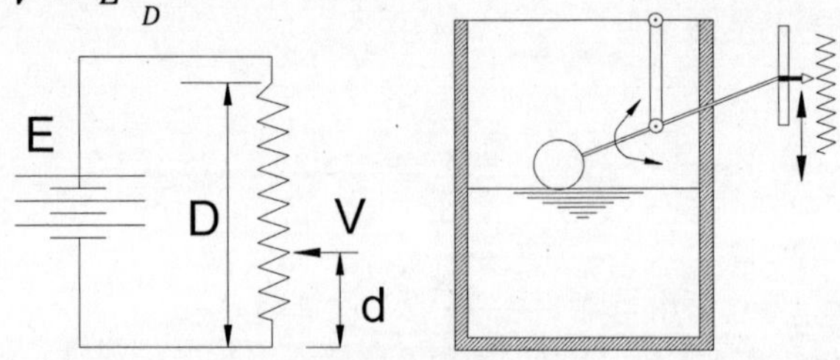

(a) Potentiometer (b) Fluid level sensor

Fig. 2.46: Use of resistance

The potentiometer type sensor may be used for water level detection as shown in Fig 2.26 (b). In this type of sensor, the output voltage, V depends upon the ratio of length, hence, the effect of temperature or environment is almost zero. The float is connected to a potentiometer wiper.

Based on this principle, there may be another type of measurement system, which is called pressure-sensitive potentiometer as shown in Fig 2.47. It consists of two strips; the upper strip is a flexible one with a metallic coating on the underside face. The bottom is a fixed strip with a coating of resistive material with resistance ranging from kilo-ohm to mega-ohm. Both the strips are connected to an electric circuit. When the upper strip is pressed at some point A, there is a change in electric path and the output voltages changes from E to V. The value of V is given by ER_x/R_0, the value depends on the length x from the left side of the sensor. The resistive layer may be deposited on polyester substrate and the pushing element may be a roller.

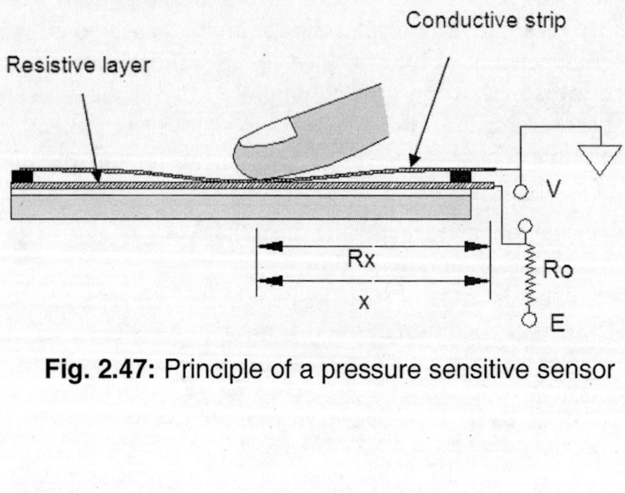

Fig. 2.47: Principle of a pressure sensitive sensor

Although, the principle of potentiometric sensor is simple, but there are some drawbacks to this type of sensors. The main disadvantages are:

a) Development of frictional force
b) Necessity of a physical coupling with the object
c) Low speed
d) Generation of heat
e) Susceptibility to dust

Another way to calculate the displacement is by using the capacitive displacement sensor. This displacement may be produced by force, pressure, temperature, etc. The principle of operation of a capacitive sensor is already discussed. Due to the displacement, there will be a variation in the capacitance which will induce a change in the electrical signal. The capacitive sensor may be monopolar i.e., only one capacitor, differential (with two capacitors) or a capacitive bridge (with four capacitors). For multi-capacitor use, one or two capacitors may be fixed. Consider a simple example with three equally spaced capacitor plates as shown in Fig 2.35. The three plates constitute two capacitors C1 and C2. If the upper and lower plates are fed with out-of-phase sine wave signals, the central plate will have no voltage, since the charges on C1 and C2 cancel each other. If due to external force the central force moves downward through a distance x, the capacitance of each capacitor will change. Following will be the changed capacitances:

$$C1 = \frac{\varepsilon A}{x_0 + x} \quad \text{and} \quad C2 = \frac{\varepsilon A}{x_0 - x} \tag{2.28}$$

There will be a change in the output voltage, which is given by

$$V_{out} = V_0\left(-\frac{x}{x_0 + x} + \frac{\Delta C_0}{C_0}\right) \tag{2.29}$$

In this case, so far $x \ll x_0$, the output voltage is a linear function of displacement.

The position and displacement can also be sensed with an electromagnetic sensor. A magnetic flux coupling between two coils may be altered by the movement of an object and subsequently converted into voltage. Variable inductance sensors use a non-magnetized ferromagnetic medium to alter the magnetic resistance of the flux path. These are known as variable reluctance transducers. It contains two coils: primary and secondary. The primary coil carries an AC excitation (V_{ref}) that induces a steady AC voltage in the secondary coil as shown in Fig 2.48. The induced amplitude depends on the flux coupling between the coils. There are two techniques to change the coupling. One is the movement of a ferromagnetic material within the flux path. This changes the reluctance (magnetic resistance) of the path. This is the basis for operating linear variable differential transducer (LVDT), rotary variable differential transformer (RVDT). The other method is to physically move one coil with respect to the other.

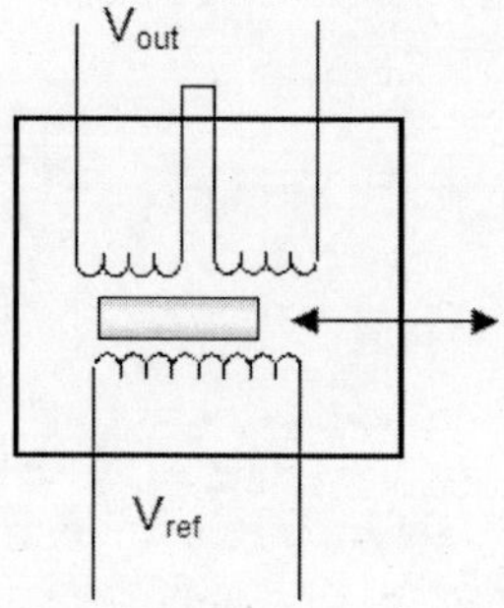

Fig. 2.48: Circuit diagram of LVDT sensor

The following are the advantages of LVDT and RVDT:

1. The sensor is a non-contact device, hence there is very little frictional resistance between the moving coils
2. Hysteresis is negligible
3. Output impedance is very low
4. Low susceptibility to noise and interferences
5. Construction is solid and robust
6. Infinitesimal resolution is possible.

Velocity :

Velocity is defined as the distance covered per unit time, or in other words, it is related to change in position over time. It is also defined as the first derivative of a position. Hence, the displacement sensors cannot be used to measure the velocity. The velocity may be linear or angular. The velocity of large objects either on land or at sea can be easily determined now using a Geo-positioning System (GPS). The location of the object is determined by receiving radio signals from a number of Earth satellites. The time delay between two signals from two satellites will give the velocity of the object. However, for small objects, the GPS system will not work. It is known that if a permanent magnet moves within a coil, a voltage is induced in the coil as per Faraday law. The voltage is proportional to the magnet velocity and the field strength. This principle is used in the velocity sensor, where a permanent magnet moves within a fixed geometry coil. The induced voltage is proportional to the relative velocity of the magnet. The full length of the magnet moves within a coil. For a single coil, the voltage induced on one end is equal and opposite to that induced on another end, hence, there is no voltage output for a singly coil sensor. To overcome this problem, the coil is divided into two sections. The north pole of the magnet induces a current in one coil, while the south pole induces a current in the other coil. The two coils are connected in a series opposite direction to obtain an output proportional to the magnet's velocity. Fig 2.49 shows a moving magnet velocity sensor. Thus the velocity sensor is similar to an LVDT sensor in operating principle with the following differences:

1. LVDT is an active sensor with a moving ferromagnetic core, whereas the velocity sensor is a passive sensor with a moving permanent magnet. In a passive sensor the current is generated within the sensor and no external current source is required.
2. Since the velocity sensor uses a moving magnet within a coil, hence the length of the sensor is a constraint, which is not a hindrance to the working of an LVDT.

Table 2.2 is a specification of a pressure sensor mentioned in (Fraden, 2003)

Table 2.3: Specification of a pressure sensor

Characteristics	Value
Magnet core displacement, in	0.5 - 24
Sensitivity (mV/in/s)	35 - 500
Coil resistance (kΩ)	2 - 45
Coil inductance (henry)	0.06 – 7.5
Frequency response (Hz) (at load > 100 coil resistance)	500 - 1500
Weight (g)	20 - 1500

In summary, the linear velocity sensor detects velocity along a distance that is limited by the size of the sensor. These sensors measure the vibration velocity. The angular version of the same sensor may measure the rotation rate continuously for any number of turns.

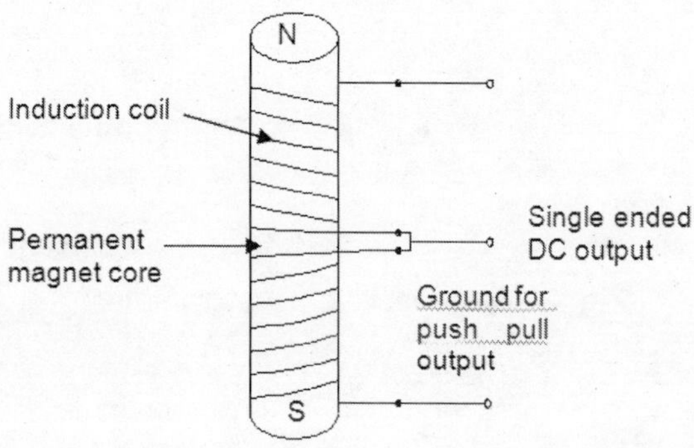

Fig. 2.49: Operating principle of an electromagnetic velocity sensor

Acceleration

Acceleration is a dynamic characteristic of an object. It requires a force for its application. Position, velocity and acceleration are correlated in the fact that velocity is the first derivative of a position and acceleration is the second derivative of a position. When an accelerometer is chosen for a particular application, the following questions are to be answered:

1. What is the anticipated magnitude of linear acceleration ?
2. What is the ambient temperature and how fast the ambient temperature may change ?

3. What is the frequency range ?
4. What is the accuracy required ?
5. What is the maximum tolerance ?
6. What kind of power supply is available?
7. What is the environmental condition ?

A capacitive sensor can be used as an accelerometer. A mass m can be attached to the central plate. The top and bottom plates are separated by distances d_1 and d_2 respectively from the central plate. The mass is supported by four silicon springs as shown in Fig 2.50. All three parts are micromachined from a silicon wafer. When the mass moves by a distance Δ towards the upper plate the distance d_1 is reduced by Δ and the distance d_2 is increased by Δ. The mechanical force F_m acting on the mass is given by

$$\Delta = \frac{F_m}{k} \tag{2.30}$$

Where, k is the spring constant of the silicon springs.

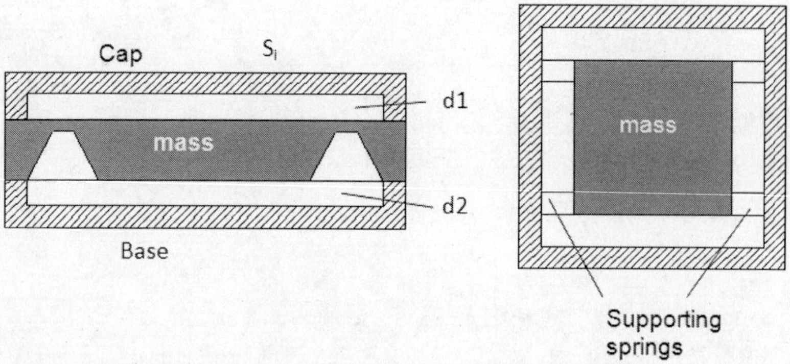

(a) Sectional elevation (b) Plan

Fig. 2.50: Capacitive accelerometer

This type of accelerometer cannot measure vibrations of large amplitude. The maximum displacement is around 20 μm.

Piezoelectric accelerometers are widely used now a days. The mechanical energy is directly converted to electrical energy in a crystalline material composed of electrical dipoles. There are many advantages of this type of accelerometer, namely, they operate on a wide range of frequency from 2 Hz to 5 kHz, and they possess goof off-axis noise reduction, high linearity, wide operating temperature range (upto 120^0 C). The commonly used piezoelectrical materials constitute quartz crystals, barium titanate, lead zirconitetitanate and lead metaniobite. The crystal is placed between the case and the seismic mass that exerts on it the force proportional to the acceleration. Fig-2.51.

One schematic diagram of the piezoelectric accelerometer is shown in Fig 2.52. Acceleration of the base moves it relative to the mass, which exerts a force on the crystal. The output is directly proportional to the acceleration.

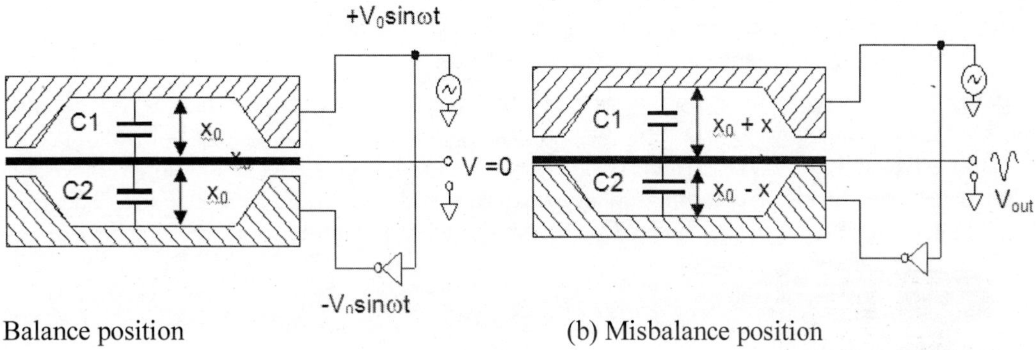

Fig. 2.51: Working of a piezoelectric accelerometer

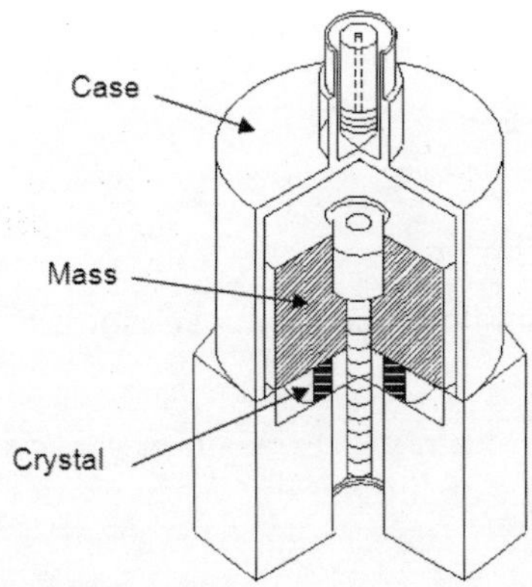

Fig. 2.52: Schematic diagram of piezoelectric accelerometer

Force

Force is defined as a product of mass and acceleration. Thus, the force is a vector quantity. Thus, when acceleration is zero, the force is also zero. The mass, m of a body depends upon the quantity of matter per unit volume called mass density. The mass density, ρ is defined as follows:

$$\rho = \frac{m}{V} \tag{2.31}$$

Where, V is the volume of the body. Thus, force is applied on a spot.

The force sensors are of two types : quantitative and qualitative. In quantitative force sensor the output is a value represented by an electrical signal. Examples are strain gauges, load cells etc. The qualitative sensor does not give any value but indicates whether any predetermined threshold value is exceeded or not. An example is the computer key board, where the pressure on the key exceeds a threshold value for its operation. A pressure sensitive floor mat and a piezoelectric cable are two other examples of qualitative force sensors. There are different ways by which an unknown force can be measured. The modern sensors utilize the following principles:

1. Measuring the strain induced in a material
2. Measuring the fluid pressure generated due to the unknown force
3. Measuring the balancing electromagnetic force.

An LVDT can be used to measure the applied force. The force sensor consists of a spring connected to an LVDT as shown in Fig 2.53. Within the elastic limit the deformation of the spring will be transferred to the LVDT which will generate an output voltage. The unknown force can also be determined using a combination of pressure sensor and fluid filled below. The fluid filled bellows will act as a converter to convert the localized force to a pressure distributed over a sensing membrane.

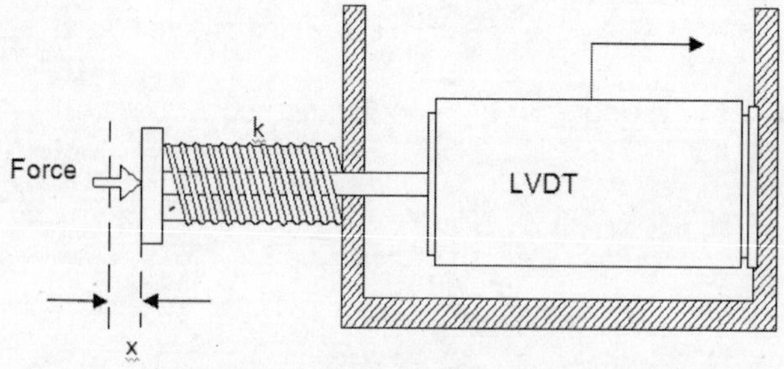

Fig. 2.53: Spring loaded force sensor with LVDT

The force can also be measured using a strain gauge. A strain gauge is an elastic resistive sensor, whose resistance is a function of the applied strain. When a force is applied to a strain gauge, there is a deformation i.e. a strain is developed. The strain, e is related to the change in resistance, dR through eq.2.31

$$\frac{dR}{R} = S_e\, e \tag{2.32}$$

Where R is the resistance prior to application of force, S_e is the gauge factor and e is the strain. For most of the materials dR is around 2.0, whereas, the dR≈ 6.0 for platinum. Other materials are Nichrome, Advance and Karma. Typical resistances vary from 100 ohms to several thousand ohms. Due to an applied force, if the strain gauge is moved by a distance, x = S_e e, the total resistance is given by

$$R = R_o(1 + x) = R_o(1 + S_e\, e) \tag{2.33}$$

The applied force can also be measured using a piezoelectric material. It is based on the principle that when an asymmetrical elastic crystal is deformed by a force, an electrical potential will be developed within the distorted crystal. This effect is reversible. It means, when an electrical potential is applied between two surfaces of a crystal, the crystal will undergo a deformation. The magnitude and polarity of a surface charge, Q is proportional to the magnitude and direction of the applied force, F ,i.e.

$$Q = k\,F \tag{2.34}$$

Where, k is the constant the crystal termed as charge sensitivity. From the Hooke's law the force F can be expressed as :

$$F = \frac{a\,Y\,\Delta t}{t} \tag{2.35}$$

Where, a is the surface area of the crystal, Y is the Young's modulus, Δt is the change in thickness and t is the original thickness of the crystal.

The charge at the electrode is given by

$$E_0 = \frac{Q}{C} = \frac{Q}{\varepsilon a/t} \tag{2.36}$$

Where, C is the capacitance of the electrodes in Farad and ε is the absolute permittivity.

$$E_0 = \frac{k\,F}{\varepsilon\,a/t} = g\,\frac{t}{a}\,F \tag{2.37}$$

Where, g is the voltage sensitivity = k/ε, volt m / N

There are different piezoelectric materials like Rochelle salt, tourmaline, quartz, ammonium dihydrogen phosphate (ADP), lithium sulphate, barium titanate and lead zirconatetitanate (PZT). The range of forces that can be measured with a piezoelectric sensor varies from 1 to 200 kN.

Piezoelectric crystals can also be used to measure any dynamic force. The thin plate of quartz will be oscillating under the external force, the oscillations are different from the natural frequency. Any distortion of the shape of the plate caused by an external force will alter the oscillation frequency. Thus the dynamic force can be measured by the change in frequency of the oscillator.

A capacitive sensor can also be used to measure an applied force. The force is directed onto the membrane of a capacitor plate whose elastic deflection is detected by a capacitance variation. This capacitance variation can be converted to a DC voltage variation.

Pressure

To understand how pressure can be measured in a sensor, it is necessary to know how pressure is applied to a matter. A matter can be solid or fluid. A fluid can flow and exists in two forms, liquid and gas. By applying pressure it is possible to convert liquid into a gas and vice versa. Again, to a fluid, pressure cannot be applied at an angle, because in this case the fluid molecules will slide. Hence to a fluid, pressure is to be applied normally i.e. at an angle of 90^0. Thus for a fluid at rest, the pressure is defined as the force applied normally to a unit area surface

$$p = \frac{dF}{dA} \tag{2.38}$$

Pressure is a scalar quantity; hence, it does not depend on direction. Pressure applied to any boundary of a confined substance is equally transferred to all other boundaries. Using the kinetic theory of gases, the pressure can be expressed as a measure of kinetic energy, KE as follows :

$$p = \frac{2}{3}\frac{KE}{V} = \frac{1}{3}\rho C^2 = NRT \tag{2.39}$$

Where, KE is the kinetic energy, V is the volume, ρ is the density, C^2 is an average value of the square of the velocity of molecules, N is the number of molecules per unit volume, R is the specific gas constant and T is the absolute temperature.

The above equation suggests that for gases, the pressure is directly proportional to the density. For example, at 0^0C and at 1 atmospheric pressure, air has a density of 1.3 kg/m^3, whereas at the same temperature but 50 atmospheric pressure, density of air is increased to 65 kg/m^3. On the other hand, the density of liquid does not increase proportionately with pressure. For example, at 0^0C and at 1 atmospheric pressure, water has a density of 1000 kg/m^3, whereas, at the same temperature but 50 atmospheric pressure, density of water is increased to 1002 kg/m^3. Moreover, the gas pressure may be expressed as overpressure or partial vacuum. An overpressure means a pressure higher than the ambient air pressure. A negative or partial vacuum means a pressure less than the ambient air pressure. A relative pressure means a pressure with respect to ambient pressure. An absolute pressure means a pressure with respect to vacuum or zero pressure.

Unit of pressure :

The SI unit of pressure is 1 Pascal, Pa which means a force of 1N acting over an area of 1 m^2. In many cases the ambient pressure is expressed as 1 atm (atmosphere). 1 atm is defined as a pressure exerted on an area of 1 cm^2 by a water column of height 1m at a temperature of 4^0C. The relation among Pa, psi (pound per square inch), atm and cm-Hg (1 cm of Mercury column) is given below :

$$1 \text{ Pa} = 1.45 \times 10^{-4} \text{ psi} = 9.869 \times 10^{-6} \text{ atm} = 7.5 \times 10^{-4} \text{ cm Hg}$$

It is noted that 1 Pa pressure is very low, with only 0.1 mm of water. There is one more unit of pressure which is normally used in the industries. It is termed as Torricelli expressed as Torr. 1 Torr is defined as the pressure exerted by 1 mm height of mercury at 0^0C under normal pressure and normal gravity. 1 atm pressure is equivalent to 760 mm of mercury, i.e., the following relationship holds good:

$$1 \text{ atm} = 760 \text{ mmHg} = 760 \text{ Torr} = 101325 \text{ Pa}$$

A pressure sensor is a complex sensor. There is more than one step by which the pressure is finally converted to an electrical signal.

A number of sensors can be used to measure the pressure. Following are a brief description of the sensors.

Mercury pressure sensor: this sensor is used to measure the pressure of a gas. A U-shaped wire is immersed into mercury as shown in Fig 2.54. The wire is connected to a Wheatstone bridge circuit. The resistance of the wire balances the bridge circuit as long as the differential pressure in the U-tube is zero. When pressure is applied, the mercury columns in the U-tube will have a differential height which will change the resistance of the wire. The higher the pressure, the more will be the resistance of the wire in that arm and less in the other arm. The difference in resistance is directly proportional to the output voltage.

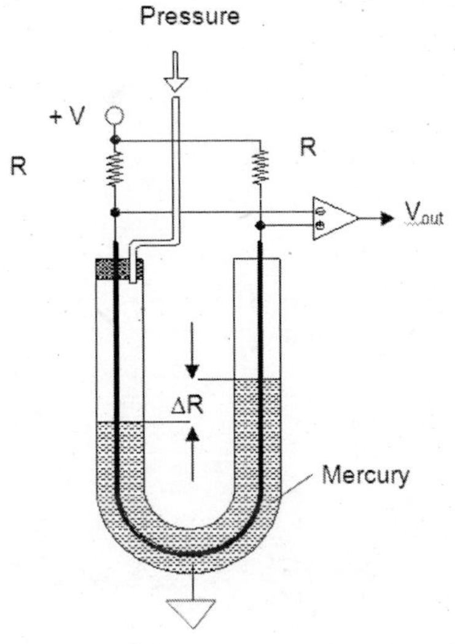

Fig. 2.54: Mercury filled pressure sensor

Membranes and thin plates:

Membranes and thin plates can be used to measure the pressure. Due to application of the pressure, the membrane will deform which can be suitably converted to a voltage using the displacement sensor. A typical membrane sensor is shown in Fig 2.55 when pressure is applied the strain gauge will deform and an electrical signal will be produced by means of piezoresistivity. Nowadays, a great majority of the pressure sensors are made of silicon membranes.

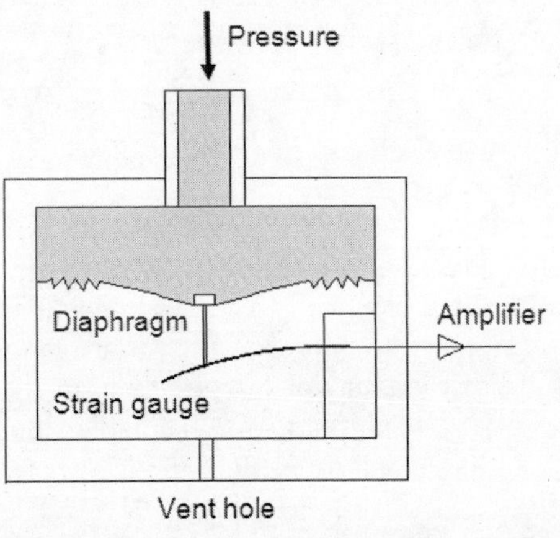

Fig. 2.55: Metal diaphragm for converting pressure to linear deflection

When pressure is applied to a membrane, a radial tension S in kN/m is developed along the edge. There will be a maximum defection, z_{max} at the centre of the membrane. The stiffness to the bending stress will be negligible if radium is more than 200 times the thickness. The central deflection z_{max} is given by

$$z_{max} = \frac{r^2 p}{4S} \quad (2.40)$$

$$\sigma_{max} \approx \frac{S}{t} \quad (2.41)$$

Where, p is the stress and t is the thickness of the membrane and σ_{max} is the stress. When the thickness of the membrane is not negligibly small, i.e. r/t is less than 100, the property of the membrane will be considered as thin plate. The maximum deflection is linearly proportional to the pressure

$$z_{max} = \frac{3\,(1-\mu^2) r^4 p}{16\,E\,t^3} \quad (2.42)$$

Where, E is the Young's modulus and μ is the Poisson's ratio. The maximum stress for the thin membrane is given by

$$\sigma_{max} \approx \frac{3\,r^2 p}{4\;t^2} \quad (2.43)$$

The bending of thin plate and membrane under normal pressure is shown in Fig 2.56

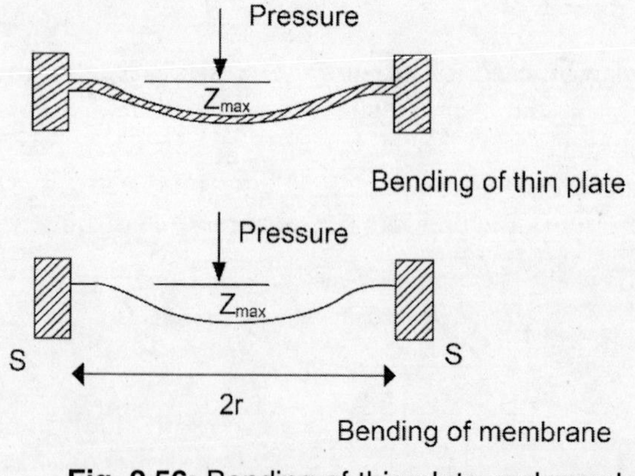

Fig. 2.56: Bending of thin plate and membrane

This deflection is now to be converted to an electric signal. This can be achieved by considering silicon membrane with impregnated piezoresistive gauge resistors made by diffusive impurities into the diaphragm. It is customary to fabricate the strain gauge resistors connected as the Wheatstone bridge. The full scale output of this circuit is in the range of millivolt, hence a signal conditioner is used to bring the output to an acceptable value. When stress is applied to a semiconductor resistor with an initial resistance, R. there is a change in the resistance ΔR [7]

$$\frac{\Delta R}{R} = \pi_l \sigma_l + \pi_t \sigma_t \quad (2.44)$$

Where, π_l and π_t are piezoresistive coefficients along longitudinal and transverse directions respectively and σ_l and σ_t are respective stresses. The π coefficients depend upon the orientation of resistors on the silicon crystals. For the positions of the resistors on the silicon diaphragm as shown in Fig 2.57 the coefficients are approximately expresses as [7] ;

$$\pi_l = -\pi_t = \frac{1}{2}\pi_{44} \tag{2.45}$$

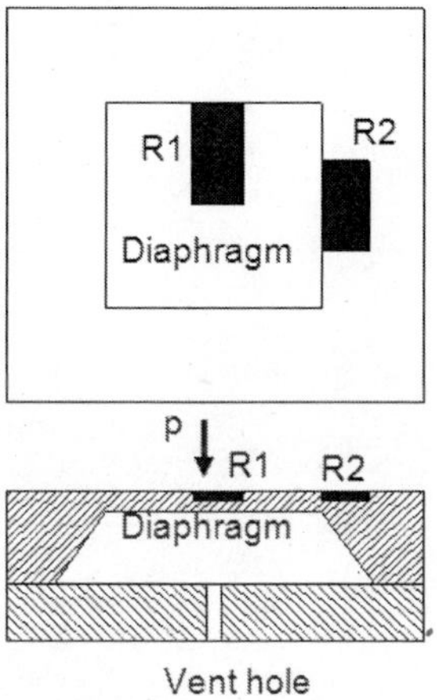

Fig. 2.57: Piezoresistor on a silicon diaphragm

Due to a change in applied pressure there is a change in applied stress, hence a change in the resistance. For the above positioning of the resistors, opposite polarity will be developed and the effective stress will be the algebraic sum of the two values.

$$\frac{\Delta R_1}{R_1} = -\frac{\Delta R_2}{R_2} = \frac{1}{2}\pi_{44}(\sigma_{1y} - \sigma_{1x}) \tag{2.46}$$

When connecting R_1 and R_2 in a half bridge circuit, and exciting the bridge with a voltage E, the output voltage is given by

$$V_{out} = \frac{1}{4}E\,\pi_{44}(\sigma_{1y} - \sigma_{1x}) \tag{2.47}$$

Capacitive sensor

The silicon diaphragm can be connected with another silicon plate to form a capacitive pressure sensor. In a capacitive pressure sensor, the displacement of the diaphragm modulates the capacitance, and thus there is a fluctuation of output voltage.

Torque sensor

The torque of a shaft can be measured with strain gauges. The strain gauges are bonded to the shaft at an angle of 45^0 with the horizontal axis of the torque shaft as shown in Fig-2.58. Typically four strain gauges are bonded and connected to a Wheatstone bridge configuration. A stable voltage,

5V dc 0r 10 V dc is required to excite the circuit. When torque will be applied, there will be a change in resistance and therefore a change in voltage.

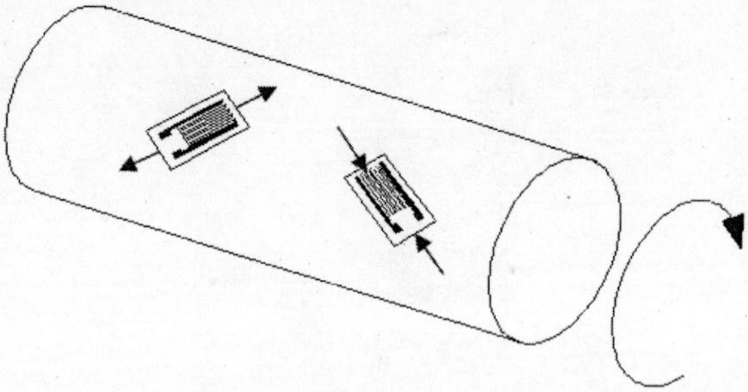

Fig. 2.58: Strain gauges bonded to a torque shaft

2.7 MEASUREMENT UNCERTAINTY

The measurement uncertainty is the maximum possible error, at a specified confidence, that may reasonably occur. Since the error is unknown, its limit, at a given confidence, must be specified. This estimate is called the uncertainty. Sometimes, the term accuracy is also used to express the maximum limit of an error. But the term accuracy is ambiguous, since it is not clear if it is stated that what is the accuracy of ±2%? Although, uncertainty is the negative statement of expected limits of an error, but the term uncertainty is unambiguous. Hence, in common practice, we use the term measurement uncertainty instead of measurement accuracy.

How to express the uncertainty of measurement

Since the true value of any measurement is uncertain, we may have two questions to be answered – (a) how large is the error and (b) how confidence we are to state that the error may fall within a given range? Thus, there are two terms to be defined for expressing the uncertainty of any measurement, they are (1) interval and (2) confidence level. Thus, if it is said that the length of a piece of rod

$$30 \text{ cm} \pm 1 \text{cm at a confidence level of } 95\%$$

It means that the length of the rod will be in the interval of 29 cm to 31 cm at a confidence level of 95%. The uncertainty of measurement is important for the following cases:

a) Calibration: while issuing a calibration certificate, the interval of uncertainty is to be mentioned.
b) Tolerance: how much tolerance we can give to a test result.
c) Test : after completing a test, whether the test result is acceptable or not

Use of basic statistics in measuring the uncertainty:

Normally it is stated that at least three measurements are to be taken for any test. The anomaly for the first two measurements is eliminated with the third measurement, since it will be then understood out of two which one is correct. However, in a production industry, there will be a large number of products of which measurements are to be taken. In such a situation, it is recommended that some statistics are to be applied. There are two statistical parameters which are required to be found out, they are

a) Arithmetic mean
b) Standard deviation.

Arithmetic mean:

The arithmetic mean is obtained by taking the average of all the observations. It is normally expressed by putting a bar over the variable. Suppose there are n number of variables of x then the arithmetic mean is expressed as

$$\bar{x} = \frac{\sum_{i=1}^{n} x_i}{n} \tag{2.48}$$

Where, x_i is the ith value of x . Many authors prefer to use the symbol μ to express the mean value.

Example 2.8

Find the arithmetic mean of the following readings of water content (%)

12.56 12.76 12.55 12.98 12.37 12.76 12.87 12.59 12.45 12.65

Ans:

Sum of 10 observations :Σw_i = 126.54 %

Arithmetic mean = 126.54/10 = 12.65 %

Normally a question is asked what is the minimum number of an observation? There is no single answer to this. The more number of readings we take, the more will be accuracy and the result will tend to the true value. Many researchers prefer minimum number is 10, since the calculation is easy. However, at least 3 observations are to be taken for taking a mean value.

Standard deviation

When the observations vary widely and the mean value may not justify the true value, it is required to find the range between the largest and the smallest values. The spread of values gives an idea of the uncertainty of the measurement. The usual way to quantify the spread is by standard deviation. It is a rule of thumb that two thirds of the observations will fall within ± one standard deviation from the average and 95% of the total observations will fall within ± two standard deviations. The formula for obtaining the standard deviation is given by:

σ= standard deviation

$$= \sqrt{\frac{\sum_{i=1}^{n}(x_i - \mu)^2}{n-1}} \tag{2.49}$$

x_i = i^{th} data value

n = number of observations.

Ex. Find the standard deviation of the following readings of water content (%)
12.56 12.76 12.55 12.98 12.37 12.76 12.87 12.59 12.45 12.65

Ans:

Sum of 10 observations : $\Sigma w_i = 126.54$ %

Arithmetic mean, $\mu = 126.54/10 = 12.65$ %

x_i	$\mu - x_i$	$(x_i - \mu)^2$
12.56	0.094	8.836×10^{-3}
12.76	-0.106	11.236×10^{-3}
12.55	0.104	10.816×10^{-3}
12.98	-0.326	106.276×10^{-3}
12.37	0.284	80.656×10^{-3}
12.76	-0.106	11.236×10^{-3}
12.87	-0.216	46.656×10^{-3}
12.59	0.064	4.096×10^{-3}
12.45	0.204	41.616×10^{-3}
12.65	0.004	0.016×10^{-3}
	Total =	0.32144

Standard deviation, $\sigma = \sqrt{(0.32144/9)} = 0.189$

Here 6 values fall within $\mu \pm \sigma$

And all values fall within $\mu \pm 2\sigma$

Example 2.9

Types of uncertainty:

There may be two types of uncertainty in a measurement. They are

Random: this type of uncertainty occurs when the measurements are not repeated. The measurement may be different or random every time. For this type of measurements, a large number of observations are to be taken and the average is considered to be true value.

Systematic : In this case, there is no new observation for a repeated measurement under a similar situation. Special methods are needed to estimate the uncertainty for systematic measurement.

Distribution or shape of error

The distribution of errors: in most of the cases the distribution of errors takes a bell shape. This is called the normal Gaussian distribution. The normal distribution or Gaussian distribution is a continuous distribution of a random variable i.e., error. The distribution is also called the probability density function. The expression of the distribution, i.e., frequency is given by:

$$y = f(x) = \frac{1}{\sigma\sqrt{2\pi}} e^{-\frac{(x-\mu)^2}{2\sigma^2}} \tag{2.50}$$

Where, x = random variable obtained from the measurement, it may be error

μ= mean value of the set of variable

σ= standard deviation as defined above.

Typically, n, μ and σ are unknown and are to be determined. The distribution curve is normally bell shaped and is shown below:

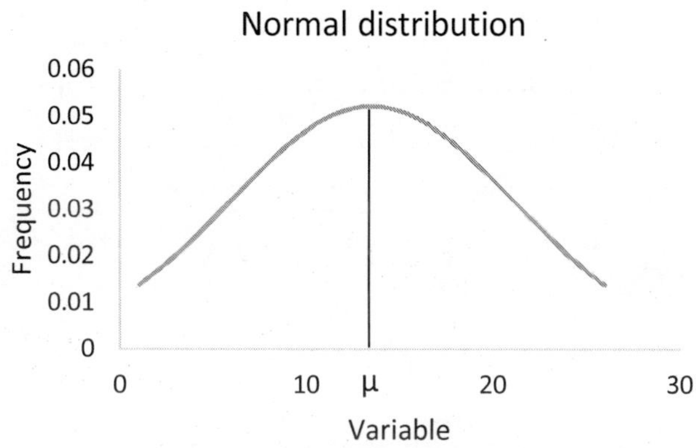

Fig. 2.59: Normal distribution

There is another type of distribution, called uniform or rectangular distribution. When the measurements are quite evenly spaced between the highest and lowest values, a uniform distribution will occur. The probability of occurrence of a value is constant in a uniform distribution. If the maximum value of an observation x is b and the lowest value is a for a measurement, the distribution function P(x) is given by the Fig ----. The probability of occurrence of any value lying between a and b is given by 1/(b-a). The mean value and the standard deviation for a uniform distribution are given below:

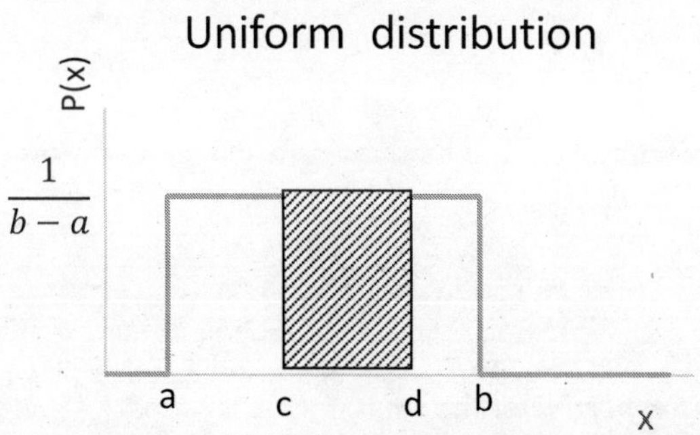

Mean value $\mu = \frac{a+b}{2}$

Standard deviation =

$$\sigma = \sqrt{\frac{(b-a)^2}{12}}$$

Fig. 2.60: Uniform distribution

Probability of occurrence of any value lying between c and d is given by

$$P(c < X < d) = \frac{d-c}{b-a}$$

Ex. In a standard proctor test of a sandy soil, it is expected that the maximum dry density will vary from 15 kN/cu.m to 20 kN/cu.m. Find the probability that the density will be more than 18 kN/cu.m.

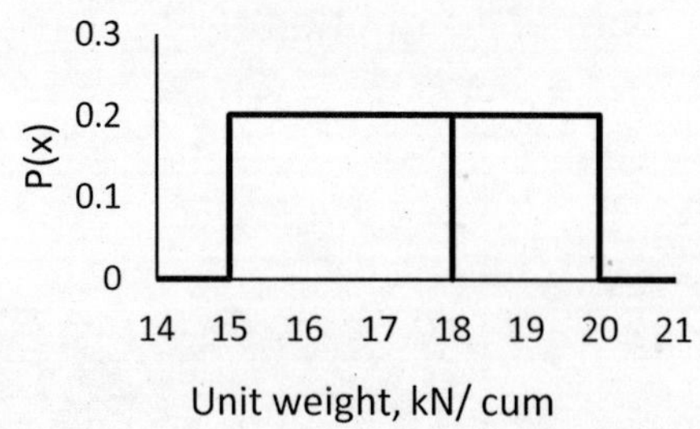

Ans

Mean value = 17.5 kN/cu.m

$$\text{Standard deviation} = \sqrt{\frac{(20-15)^2}{12}} = 1.44 \ kN/cum$$

$$P(18 < X < 20) = \frac{20-18}{20-15} = 0.4$$

Example 2.10

Sometimes we confuse human mistake as an uncertainty to a measurement. But human mistake is not an uncertainty. Following jobs are not measurement uncertainty:

a) Human mistake
b) Tolerance of a job
c) Specification of an item
d) Accuracy or inaccuracy of a job. Accuracy is a qualitative term whereas, uncertainty is a quantitative term.
e) Error

Estimation of uncertainties

There are two ways to estimate an uncertainty irrespective of the source. One is "type A" and the other is "type B". In some measurements both the types are used. Type A evaluation is carried out using statistics over repeated readings. Type B is obtained from any information like past experience, calibration certificate, manufacturer's specifications, published information, common sense, etc.

There are mainly eight steps to be followed to obtain the measurement uncertainty. They are:

(i) Decide what is to be found out from the measurements. This includes choice of actual measurements and calculations
(ii) Carry out the measurements
(iii) Estimate the uncertainty of each input. All uncertainties are to be expressed similarly
(iv) Decide whether the errors of the input quantities are independent or not.
(v) Calculate the measurement if any correction is known from calibrations
(vi) Find the combined standard uncertainty from all the individual inputs.
(vii) Express the uncertainty in terms of a coverage factor including the uncertainty interval and level of confidence.
(viii) Write down the measurement result and the uncertainty.

Calculation of standard uncertainty:

All calculating uncertainties will be expressed at the same confidence level by converting them to standard uncertainty. At first, the mean μ and the standard deviation of σ all the measurements (Type A) are obtained. The standard uncertainty, u is then obtained as

$$u = \frac{\sigma}{\sqrt{n}} \tag{2.51}$$

Where, n is the number of observations. The standard uncertainty is also called the standard of the mean or standard error of the mean.

Example 2.11

Ex. Estimate the value of a quantity w and repeatability for the value from the following four measurement :

12.56 12.76 12.55 and 12.98

Ans:

The mean value = 12.712

Standard deviation = 0.203

The estimated repeatability = $\frac{\sigma}{\sqrt{n}} = \frac{0.203}{\sqrt{4}} = 0.101$

For type B measurements, where the information is more scarce, an estimate of the upper and lower limits is to be ascertained. It will be assumed that the value will lie equally in the range. Hence, for such a case, a uniform distribution is assumed. The standard uncertainty for a uniform distribution is given by

$$\frac{a}{\sqrt{3}}$$

Where, a is the semi-range or half-width between the upper and lower limits.

Finally, all uncertainties are to be combined to find the total uncertainty. To achieve this, all the uncertainties are to be expressed in a common unit. This is required because a temperature uncertainty cannot be added to a length uncertainty. The temperature uncertainty is to be converted to the length uncertainty. This can be obtained by knowing the effect of temperature on the length of the material. If it is known that there will be a change of 0.1 percent in length for every degree rise in temperature, the temperature uncertainty can be converted into the length uncertainty. For an example, under this case, a temperature uncertainty of $\pm 2^0$ C is equivalent to a length uncertainty of ± 2 cm for length of 100 cm.

The standard uncertainties with the same unit may be combined by adding. Suppose a long fencing has several panels and the length uncertainty of the panels are a, b, c, ..., then the total uncertainty is given by

$$\text{Combined uncertainty} = \sqrt{a^2 + b^2 + c^2 + \cdots}$$

For a measurement that is obtained by multiplication or division of some constituent measurements, a relative uncertainty is obtained instead of the addition of uncertainties. For example, suppose the uncertainty of an area measurement is to be found out from the measurements of length, L and breadth, B, i.e., area A = length L x breadth B. If we consider the uncertainty in length measurement is u(L) and in breadth measurement is u(B), the relative uncertainties are u(L)/L and u(B)/B respectively. The relative uncertainty of the area measurement is given by:

$$\frac{u(A)}{A} = \sqrt{\left(\frac{u(L)}{L}\right)^2 + \left(\frac{u(B)}{B}\right)^2} \tag{2.52}$$

In case the measurement consists of the product of three sub-measurements, for example, volume = length x breadth x height, the same procedure will be followed, except that instead of two terms there will be three terms. The same procedure as above will be followed if the result is the quotient of two values i.e., one value is divided by the other value.

For a squared value, i.e., S^2 (say), the relative uncertainty is given by

$$\frac{2\,u(S)}{S} \tag{2.53}$$

For a square root value, i.e., $\sqrt{S}$ (say), the relative uncertainty is given by

$$\frac{u(S)}{2S} \tag{2.54}$$

When the measurement is a composite, the relative uncertainties are obtained using the above expressions. For example, the electric power is expressed as

$$P = \frac{V^2}{R} \tag{2.55}$$

Where P is the power, V is the voltage and R is the resistance. The relative uncertainty of P is obtained as

$$\frac{u(P)}{P} = \sqrt{\left(\frac{2\,u(V)}{V}\right)^2 + \left(\frac{u(R)}{R}\right)^2} \tag{2.56}$$

The combined uncertainty may be considered as 'one standard deviation'. This can be further scaled to some confidence level, say 95%. This can be done by multiplying the combined uncertainty with a coverage factor k. When the combined uncertainty, u_c is multiplied by a coverage factor k, the resulting uncertainty is called an expanded uncertainty U and is given by

$$U = ku_c \tag{2.57}$$

Most commonly, the value of k is taken as 2 for a confidence level of 95% (for normally distributed). Some coverage factors are mentioned below (applicable only for normal distribution)

k = 1 for a confidence level of 68%

= 2 for a confidence level of 95%

= 2.58 for a confidence level of 99%

= 3 for confidence level of 99.7%

Ex. Calculate the uncertainty in the length of a piece of string

Ans:

To solve this problem, it is needed first to ascertain the sources of error which may lead to the uncertainty. There are three broad sources of error:

1. Possible errors during measurement with the tape
    - Calibration uncertainty
    - Error due to stretching
    - Error due to temperature susceptibility
    - Error due to permanent bending effect in the roll.
    - Resolutions of the tape i.e. what is the minimum division on the tape
2. Possible errors with the string
    - Whether the string is overstressed or under stressed
    - Any temperature effect on the string
    - Any kinks at the beginning or end of the string
3. Possible errors due to the measuring process.
    - Whether the tape is laid parallel to the string
    - Whether the zero of the tape exactly coincides with the starting point of the string
    - How much repeatability is the measurement

Step 2: Carry out the measurements

Suppose there are 10 measurements with a 30 m tape. The following are the two parameters obtained from the observations:

Mean : 15.012 m

Standard deviation : 0.0025 m

Step 3: Obtain the standard uncertainty

It is a combination of the following uncertainties:

a) Uncertainty from calibration certificate. The calibration certificate normally contains the expanded uncertainty. Consider the calibration uncertainty is 0.1 percent with a coverage factor k =2 (for a normal distribution). For 0.1 percent uncertainty the value is roughly 15 mm. Thus the standard uncertainty is $u1 = 15/2 = 7.5$ mm
b) Uncertainty about reading. The least value which can be measured on a tape is 1mm. This is a uniform distribution. Hence, the standard uncertainty for reading = half width divided by $\sqrt{3}$. The standard uncertainty = $0.5/\sqrt{3} = 0.3$ mm

Example 2.12

c) Uncertainty on measured length. The tape may have some bends due to which the measured length will be less than the actual length. It is normally considered that the total uncertainty due to this underestimation is 0.2%. Thus, the total length is to be increased by 0.2% of 15.012 m = 30 mm. The corrected length = 15.012 + 0.03 = 15.042m. This is a uniform distribution case and the standard uncertainty will be half-width divided by √3. i.e., standard uncertainty = 15/√3 = 8.6 mm

All the above uncertainties are type B uncertainty.

Standard uncertainty for type A measurement

$$= \frac{\sigma}{\sqrt{n}} = \frac{2.5}{\sqrt{10}} = 0.8\ mm$$

Step 4: Check whether all the input quantities are independent or not. If not, then some more calculations are to be added. Consider that all the input quantities are independent.

Step 5: Calculate the corrected value of the measurement

= 15.012 + 0.03 = 15.042 m

Step 6: Combined standard uncertainty = $\sqrt{7.5^2 + 0.3^2 + 8.6^2 + 0.8^2}$ = 11.4 mm

Step 7: express the uncertainty in coverage factor:

Consider the coverage factor k =2 for a 95% confidence level. As such, the final corrected standard uncertainty = 22.8 mm.

Step 8 : final expression. The final expression can be written as 15.042 ± 0.228 m

The above uncertainty is obtained with a coverage factor of 2 and a 95% confidence level.

Example 2.13

Analysis of uncertainty – spreadsheet model.

The above example can be expressed in a spreadsheet model as shown below

Source of uncertainty	Value ±	Probability distribution	Divisor	Standard uncertainty
Calibration uncertainty	15 mm	Normal	2	7.5 mm
Resolution	0.5 mm	Rectangular	√3	0.3 mm
String not lying straight	15 mm	Rectangular	√3	8.6 mm
Standard uncertainty	0.8 mm	Normal	1	0.8 mm
Combined standard uncertainty		Assumed normal		11.4 mm
Expanded uncertainty		Assumed normal		22.8 mm

UNIT SUMMARY

This unit discusses in short different types of sensors that are usually used in civil engineering applications. The main applications include the measurement of strain, displacement, velocity, acceleration, force, pressure and torque. The basic concept for measurement of all these physical quantities is discussed. The selection of appropriate sensors is also discussed. Once selected the sensors are to be properly installed; the installation methods are also discussed. Finally, a short discussion on measurement uncertainty is also incorporated. The calibration of instruments is an important part of an accurate measurement; the calibration technique is also discussed. An attempt has been made so that the students can fabricate their instruments

EXERCISES

Example 2.14

Ex. Obtain the Poisson's ratio of a resistance wire strain gauge using a soft iron wire of small diameter with a gauge factor 4.0. Neglect the piezo-resistive effect.

Ans

Given, gauge factor = 4.0

Piezo resistive effect is neglected

Piezo resistive effect is neglected

Thus, $GF = (1 + 2\gamma)$

$4.0 = 1 + 2\gamma$

$\gamma = 1.5$

Ex. A strain of 10 microstrain is caused in a structural member when subjected to a compressive force. Two separate strain gauges are attached to the structural member, one is a nickel wire strain gauge with gauge factor = -12.0 and the other is nichrome wire strain gauge (gauge factor =2). If the resistance of strain gauges before being strained is 105 Ω, calculate the change in the value of resistance of the gauges after they are strained.

Example 2.15

Ans

Given, strain = -10×10^{-6}, the negative sign is due to compression

Gauge factor for nickel = -12.0

Gauge factor for nichrome = 2.0

The initial resistance of the gauge = 105 Ω

We know that gauge factor $GF = \frac{\frac{\Delta R}{R}}{\epsilon}$

For nickel wire $\frac{\Delta R}{R} = -12 * (-10 * 10^{-6}) = 120 * 10^{-6}$

$\Delta R = 105 \times 120 \times 10^{-6} = 12.6 \times 10^{-3} = 12.6\ m\Omega$

Thus, there is an increase in resistance by 12.6 mΩ

For nichrome wire $\frac{\Delta R}{R} = 2 * (-10 * 10^{-6}) = -20 * 10^{-6}$

$\Delta R = 105 \times (-20 \times 10^{-6}) = -2.1 \times 10^{-3} = -2.1\ m\Omega$

Thus, there is a decrease in resistance by 2.1mΩ

Ex. A simple electrical strain gauge of resistance 150 Ω and having a gauge factor of 2 is bonded to a steel having an elastic limit stress of 390 MN/m^2 and a modulus of elasticity of 210 GN/m^2. Calculate the change in resistance due to

a) Change in stress equal to 10 percent of the elastic range
b) Change in temperature of 20^{0}C, take coefficient of thermal expansion = 25 x 10^{-6} /^{0}C

Ans:

Given, R = 150 Ω , G.F. = 2

Elastic limit stress = 390 MN/m^2 ,

Modulus of elasticity E = 210 GN/m^2 = 210 x 10^9 N/m^2

a) Strain = stress/E $= \frac{0.1 * 390 * 10^6}{210 * 10^9} = 185.71 * 10^{-6}$

Now, GF = $\frac{\frac{\Delta R}{R}}{\epsilon}$

$2 = \frac{\frac{\Delta R}{150}}{185.71 * 10^{-6}}$

ΔR = 55.713 x 10^{-3} = 55.713 m Ω

b) Increase in resistance due to rise in temperature

$$R_{t2} = R_{t1}(1 + \alpha_0(t_2 - t_1))$$

$$= R_{t1}(1 + 25 * 10^{-6} * (20))$$

$$\Delta R = R_{t2} - R_{t1} = \alpha_0(t_2 - t_1) = 25 * 10^{-6} * (20) = 500\ \mu\Omega$$

Example 2.16

Ex. A strain gauge is bonded to a beam that is 15 cm long and has a cross-sectional area of 4.0 cm^2. The unstrained resistance and the gauge factor of the strain gauge are 250 Ω and 2.0 respectively. On the application of external load, the resistance of the gauge changes by 0.012 Ω. If the modulus elasticity of the steel is 210 GPa, calculate :

(i) The change in length of the steel beam

(ii) The amount of force applied to the beam

Ans :

Given, length of the beam = 15 cm = 0.15 m

Cross-sectional area = 4.0 cm^2 = 4 x 10^{-4} m^2

Unstrained resistance R = 250 Ω

Gauge Factor, GF = 2.0

$\Delta R = 0.012\ \Omega$, $E = 210\ GPa = 210 \times 10^9 N/m^2$

Now, gauge factor = $GF = \frac{\frac{\Delta R}{R}}{\epsilon}$

$$2.0 = \frac{0.012/250}{\epsilon}, \quad \text{thus } \epsilon = 24 * 10^{-6}$$

From Hook's law stress = $\sigma = E \times \epsilon = 210 \times 10^9 \times 24 \times 10^{-6}$

$= 5040 \times 10^3 N/m^2$

Force = $\sigma \times A = 5040 \times 10^3 \times 4 \times 10^{-4} = 2016$ N

Example 2.17

Ex. A single strain gauge having resistance of 150 Ω is mounted on a steel cantilever beam at a distance of 0.15 m from the free end. The beam dimensions are 30 cm in length x 2.5 cm in width and 1.0 cm in depth. An unknown force F is applied at the free end produces a deflection of 12 mm of the free end. If the change in gauge resistance is found to be 0.15 Ω, calculate the gauge factor. Take Young's modulus of steel as 200 GPa.

Ans:

Given, length of the beam = 30 cm = 0.30 m

Length of beam = 0.3 m, width of beam = 2.5 cm , depth of beam = 1.0 cm

Position of strain gauge from the free end = 0.15 m

Unstrained resistance R = 150 Ω

Change in resistance = ΔR = 0.15 Ω

Deflection at the free end = 12 mm = 0.012 m $= \frac{PL^3}{3EI} = \frac{P * 0.3^3}{3 * 200 * 10^9 * \frac{.025 * (.01)^3}{12}}$

P = 555.55 N

Bending moment at the strain location = P x = 555.55 x 0.15 = 83.33 N-m

Bending stress at the strain location $= \sigma = \frac{M}{I} y = \frac{83.33}{\frac{.025 * (.01)^3}{12}} * \frac{0.01}{2}$

= 199.99 MN/m^2

Strain $= \frac{\sigma}{E} = \frac{199.99 * 10^6}{200 * 10^9} = 1 \times 10^{-3}$

Thus, gauge factor = GF $= \frac{\frac{\Delta R}{R}}{\epsilon} = \frac{0.15/150}{1 * 10^{-3}} = 1.0$

Example 2.18

Ex. The output of a LVDT is connected to a 5V voltmeter through an amplifier whose amplification factor is 400. An output of 2.2 mV appears across the terminals of LVDT when the core moves through a distance of 0.8 mm. If the millivoltmeter scale has 100 divisions and the scale can be read to 1/4 of a division, calculate:

(ix) The sensitivity of LVDT

ii) The resolution of the instrument in mm.

Example 2.19

Ans:

(i) The sensitivity of LVDT = $\frac{output\ voltage}{displacement} = \frac{2.2}{0.8} = 2.75\ mV/mm$

(ii) Resolution of the instrument :

Sensitivity of measurement = amplification factor x sensitivity of LVDT

= 400 x 2.75 = 1100 mV/mm

1 scale division = $\frac{5}{100} = 50\ mV$

Minimum voltage that can be read on the voltmeter = $\frac{50}{4} = 12.5\ mV$

Resolution of the instrument = 12.5 x $\frac{1}{1100} = 0.011\ mm$

Ex. A capacitive transducer uses two quartz diaphragms of area 600 mm^2 separated by a distance 4.5 mm. A pressure of 500 kN/m^2 when applied to the top diaphragm produces a deflection of 0.6 mm. The capacitance is 420 pF when no pressure is applied to the diaphragms. Determine the value of capacitance after the application of pressure of 500 kN/m^2.

Example 2.20

Ans.:

Given, A= 600 mm^2, d_1 = 4.5 mm, d_2 = 4.5 – 0.6 = 3.9 mm

C_1 = 420 pF, p = 500 kN/m^2

Let C_2 = the value of capacitance after application of pressure

d_2 = distance between two plates after application of pressure

$C_1 = \frac{\epsilon_0 A}{d_1}$ and $C_2 = \frac{\epsilon_0 A}{d_2}$

$$\frac{C_2}{C_1} = \frac{d_1}{d_2} = \frac{4.5}{3.9} = 1.15$$

C_2 = 420 x 1.15 = 483 pF

Ex. A parallel plate capacitive transducer uses plates of area 400 mm^2 which are separated by a distance of 1.0 mm.

i) Determine the value of capacitance when the dielectric is air having a permittivity of 8.80 x 10^{-12} F/m

ii) Determine the change in capacitance if a linear displacement reduces the distance between the plates to 0.6 mm. Also determine the ratio of per unit change of capacitance to per unit change of displacement

iii) If a mica sheet of 0.1 mm thick is inserted in the gap, calculate the value of original capacitance and change in capacitance for the same displacement. Also calculate the ratio of per unit change in capacitance to per unit change in displacement. The dielectric constant of mica is 8.

Ans:

i) Value of capacitance $C = \frac{\epsilon_0 \epsilon_r A}{d} = \frac{\epsilon_0 A}{d} = \frac{8.80 * 10^{-12} * 400 * 10^{-6}}{1 * 10^{-3}} = 3.52$ pF

ii) Change in capacitance :

Change in displacement = 1.0-0.6 = 0.4 mm

Capacitance after displacement : $C + \Delta C = \frac{8.80 * 10^{-12} * 400 * 10^{-6}}{0.6 * 10^{-3}} = 5.867$ pF

ratio of per unit change of capacitance to per unit change of displacement

$$= \frac{\Delta C / C}{\Delta d / d} = \frac{(5.867 - 3.52)/3.52}{0.4/1} = 1.667$$

iii) When mica sheet is inserted:

The mica sheet of 0.1 mm thick is inserted into the capacitor plates with spacing = 1.0 mm

The gap between the plates now consists of two materials:

Air of total thickness = 0.9 mm

Mica sheet of thickness = 0.1 mm

$$\text{Total capacitance } C = \frac{\epsilon_0 A}{\frac{d_1}{\epsilon_1} + \frac{d_2}{\epsilon_2}} = \frac{8.80 * 10^{-12} * 400 * 10^{-6}}{(\frac{0.9}{1} + \frac{0.1}{8}) * 10^{-3}} = 3.86 \text{ pF}$$

Change in displacement = 0.4 mm

Thickness of mica = 0.1 mm and thickness of air = 0.5 mm

$$\text{Total capacitance } = \frac{8.80 * 10^{-12} * 400 * 10^{-6}}{(\frac{0.5}{1} + \frac{0.1}{8}) * 10^{-3}} = 6.868 \text{ pF}$$

Change in capacitance = 6.868 – 3.52 = 3.348 pF

$$\text{Ratio} = = \frac{3.348/3.52}{0.4/1} = 2.378$$

Example 2.21

Ex. A shaft running at a constant speed of 1500 rpm transmits a maximum power of 50 kW. Measurements of torque are made by a pair of strain gauges which are bonded on a specially machined portion of the shaft. Each gauge has a nominal resistance of 120 Ω, and a gauge factor of 2.2 and are connected electrically to the two arms of a half-activated Wheatstone bridge circuit which is energized with an excitation voltage of 8 V. The gauges have a maximum strain of 0.15%. The shear modulus of elasticity of the shaft material is 180 GPa. Calculate the following :

ix) The diameter of the shaft

ii) the output voltage and the sensitivity of the measuring system

Ans.:

Given, N= 1500 rpm, P = 50 kW, resistance of each strain gauge = 120 Ω

GF = 2.2, V = 8V, strain ∈ = 0.0015 G = 180 GPa

i) Diameter of the shaft

Power $P = \frac{2\pi NT}{60}$

$50 \times 1000 = \frac{2\pi * 1500 * T}{60}$, we get T = 318.31 N-m

We know $G = \frac{\tau}{\epsilon}$ i.e. $\tau = G\epsilon = 180 \times 10^9 \times 0.0015$

$= 270 \times 10^6$ N/m^2

Using the relation :

$$\frac{T}{I_p} = \frac{\tau}{r}$$

$\frac{318.31}{\frac{\pi}{32}d^4} = \frac{270*10^6}{\frac{d}{2}} \Rightarrow d = (18.86 * 10^{-6})^{\frac{1}{3}} = 2.66 \times 10^{-2}$ m

= 26.6 mm

ii) Output voltage and sensitivity

$GF = \frac{\frac{\Delta R}{R}}{\in}$, $\frac{\Delta R}{R} = GF * \in = 2.2 \times 0.0015 = 0.0033$

The output voltage of a half-activated bridge

$dv_0 = 2\left(\frac{\Delta R}{R} * \frac{v_0}{4}\right) = 2 \times (0.0033 \times \frac{8}{4}) = 0.0132 = 13.2$ mV

Overall sensitivity of the measuring system $= \frac{13.2}{50} =$ 0.264 mV/kW

Example 2.22

MULTIPLE CHOICE QUESTIONS

Q.1 : LVDT is a

a) capacitive transducer b) inductive transducer

c) resistive transducer d) none of them

Q.2 : LVDT windings are wound on

a) steel sheets b) aluminium

c) iron d) copper

Q.3 : Piezoelectric crystals are used for measuring changes in

a) static condition b) none of the above

c) both static and dynamic d) dynamic condition

Q.4 : Piezoelectric crystals produce emf

a) when external mechanical force is applied b) when heat is applied

c) when radio waves stimulate the crystal d) when the external magnetic field is applied

Q 5: Piezoelectric crystals are used for the measurement of

a) sound b) velocity

c) Temperature d) none of the above

Q.6 : Capacitive transducers are normally used for the following measurement

a) static b) dynamic

c) both static and dynamic d) transient

Q.7 : The gauge factor of a strain gauge is given by:

a) $GF = \frac{\Delta l/l}{\Delta R/R}$ b) $GF = \frac{\Delta l/l}{\Delta d/d}$

c) $GF = \frac{\Delta l/R}{\Delta R/l}$ d) $GF = \frac{\Delta R/R}{\Delta l/l}$

Q.8 : The gauge factor and the Poisson's ratio are related with

a) $GF = 2 + \mu$ b) $GF = 2\mu$

c) $GF = 1 + 2\mu$ d) $GF = (1 + \mu)/2$

Q.9: LVDT can be used for

a) vibration measurement b) force measurement in a beam

c) angular velocity measurement d) load measurement in a column

Q.10: an LVDT has an output in the form of

a) linear displacement of the core b) pulse

c) rotary movement of core d) none of the above

Q.11 A strain gauge should have low

a) resistant temperature coefficient b) resistance

c) gauge factor d) all of the above

Q.12 Seismic transducer is used for the measurement of

a) Linear velocity b) acceleration

c) angular velocity d) pressure

Q.13 A load cell is essentially a

a) inductive transducer b) thermister

c) resistive potentiometer d) strain gauge

Q.14 A load cell is an electro-mechanical device and is widely used for measurement of

a) static forces b) dynamic forces

c) temperature d) both (a) and (b)

Q.15 Which of the following can be used for pressure measurement

a) thermometer b) pyrometer

c) piezoelectric crystal d) bolometer

Q.16 A Wheatstone bridge circuit using strain gauges can be used for measuring

a) static strains b) dynamic strains

c) both (a) and (b) d) none of the above

Answers of Multiple Choice Questions
1(b), 2(c), 3(d), 4(a), 5(c), 6(c), 7(d), 8(c), 9(b), 10(a), 11(a), 12(b), 13(d), 14(d), 15(c), 16(c)

SHORT AND LONG ANSWER TYPE QUESTIONS

Q1. Explain briefly the terms: accuracy, errors and correction.

Q2. Discuss the working principle of LVDT

Q3. How does strain gauge load cell work ?

Q4. How does the torque sensor work ?

Q5. Discuss briefly the working of an accelerometer

Q6. What are the different types of piezometer ?

Q7. Explain the working principle of the pneumatic piezometers.

Q8. How earth pressure can be measured ?

Q9. Discuss different aspects of the installation of sensors.

Q10. What are the different types of sensors ?

Q11. Discuss the working principle of piezoelectric sensor

Q12. How does a capacitive sensor work ?

Q13. How are uncertainties estimated ?

Q14. How does an extensometer work ?

Q15. What is a load cell ?

Q16. Discuss how to measure the gauge factor for a half-bridge Wheatstone bridge circuit ?

Q17. Obtain the change in output voltage for a full bridge Wheatstone bridge circuit.

REFERENCES AND SUGGESTED READINGS

1. Alan S Morris (2001). Measurement and Instrumentation Principles, 3rd/ed, Butterworth Hienemann.
2. Beeby, S., Ensell, G., Kraft, M., and White, N (2004). MEMS Mechanical Sensors. Artech House, Inc., Boston, London
3. David A. Bell (2007). Electronic Instrumentation and Measurements 2nd /ed, Oxford Press.
4. Dunnicliff, J. (1982). Geotechnical Instrumentation for monitoring field performance, Report 89. Transportation Research Board, National Research Council, Washington, D.C
5. Fraden, J. (2003). Handbook of modern sensors. Springer-Verlag, New York.
6. https://www.automatizari-cladiri.ro/fise-tehnice/A6V10366334.pdf
7. Ilya Gertsbakh (2010). Measurement Theory for Engineers, Springer
8. Internet sources
9. Judy, J.W., R.S. Muller,and H.H. Zappe, "Magnetic Microactuation of Polysilicon Flexure Structures," Tech. Dig. Solid State Sensor and Actuator Workshop, Hilton Head, SC, 1994, pp 43-48
10. Rajput, R. K. (2017). "Mechanical Measurements and Instrumentation" .S.K. Kataria & Sons, New Delhi.
11. S. Tumanski (2006). Principle of Electrical Measurement, Taylor & Francis
12. Tanigawa H., Ishihara T., Hirata M., Suzuki K. (1985). MOS integrated silicon pressure sensor IEEE Trans Electron Devices ED -32 (7) : 1191-1195

KNOW MORE	
For more information related to this topic scan the QR code. **OR** ***Type this link in your browser*** https://digimat.in/nptel/courses/video/108105064/L34.html	

3 Data Analysis and Interpretation

UNIT SPECIFICS

Through this unit we have discussed the following aspects:

- *Fundamentals of Statistical aspects.*
- *Data reduction and interpretation.*
- *Installation and working principle of piezometer, inclinometer and strain gauges.*
- *Time domain signal processing.*
- *Discrete signals and noise.*
- *Some examples of Statistical information.*

The practical applications of the topics are discussed to generate further curiosity and creativity as well as improving problem-solving capacity.

Besides giving a large number of multiple-choice questions as well as questions of short and long answer types marked in two categories following lower and higher order of Bloom's taxonomy, assignments through several numerical problems, a list of references and suggested readings are given in the unit so that one can go through them for practice.

RATIONALE

This fundamental unit on data analysis and interpretation helps students to get a primary idea about the basic concepts on collecting field data and their interpretation through statistical information. It also discusses how a regression equation is obtained from a data source and the most probable prediction is given for a sample. The field data collected from the piezometer and inclinometer are also discussed. The types of time domain signals are discussed. A detailed review on discrete time signals is mentioned. Noise contamination of signals and procedures to remove the noise are discussed. Finally, examples are cited to obtain the mean value, mode, range, variance, and standard deviation from a given set of data.

PRE-REQUISITES

Basic Electrical and Electronic Measurements

Physics: Engineering Physics (First year, B.Tech)

Basic Statistics

UNIT OUTCOMES

The list of outcomes of this unit is as follows:

U3-O1: To predict the response of a sensor for a given sample

U3-O2: To interpret data from an uncalibrated data source

U3-O3: To reduce field data obtained from piezometer and inclinometer

U3-O4: To differentiate different types of time domain signals

U3-O5: To apply basic statistics for analyzing the data

Unit-3 Outcomes	EXPECTED MAPPING WITH COURSE OUTCOMES *(1- Weak Correlation; 2- Medium correlation; 3- Strong Correlation)*							
	CO-1	CO-2	CO-3	CO-4	CO-5	CO-6	CO-7	CO-8
U3-O1	1	1	1	1	1	3	3	1
U3-O2	1	3	1	-	1	-	2	1
U3-O3	2	1	2	-	-	1	1	1
U3-O4	-	-	-	-	2	2	1	1
U3-O5	3	-	-	1	-	1	2	1

3.1 FUNDAMENTAL STATISTICAL CONCEPTS

The term 'statistics' is obtained from the Italian word 'statista'. Statista refers to the person who deals with state-related affairs and activities. Initially, it was called 'state arithmetic' where the information regarding the nation is stated e.g. tax-related information, war plans etc. in tabular form. Statistics was initially applied in Government-related activities and data like the census. But, afterward, it was used in different fields like psychology, management, engineering and so on.

The field 'Statistics' can be described as the sub-branch of mathematics which mainly deals with the organisation, analysis and interpretation of group of numbers. Alternatively, statistics can be referred to as the science of classifying, organizing, and analysing data. King and Minium. Statics can also be reported as a mathematical science about the collection, analysis, interpretation and presentation of data.

3.1.1 Classification of Statistics

The classification of statistics is based on the extent of analysis, resulting in two main categories:

a) Descriptive Statistics and
b) Inferential Statistics.

Descriptive Statistics involves methods for organizing and summarizing large sets of data, aiming to capture the essential aspects of the information. Its focus is solely on depicting the characteristics of the collected data without delving into extrapolation. This segment is solely concerned with describing the gathered data without attempting to conclude beyond what the data itself presents. The methodology of descriptive statistics encompasses techniques for arranging (classification, tabulation, Frequency Distributions) and illustrating (Graphical and Diagrammatic Presentation) data, as well as calculations of significant data indicators such as Measures of Central Tendency and Measures of Dispersion (Variation) that encapsulate key features of the data. The details are presented as a flow chart (Fig. 3.1).

Inferential (Inductive) Statistics encompasses methods employed to deduce insights about a population based on a sample. It revolves around making statistically sound inferences about population characteristics using data derived from a sample. In this approach to statistical analysis, descriptive statistics collaborates with probability theory to generalize sample outcomes to the entire population. Hypothesis testing, exploring relationships between variables, and making predictions also fall under the domain of inferential statistics.

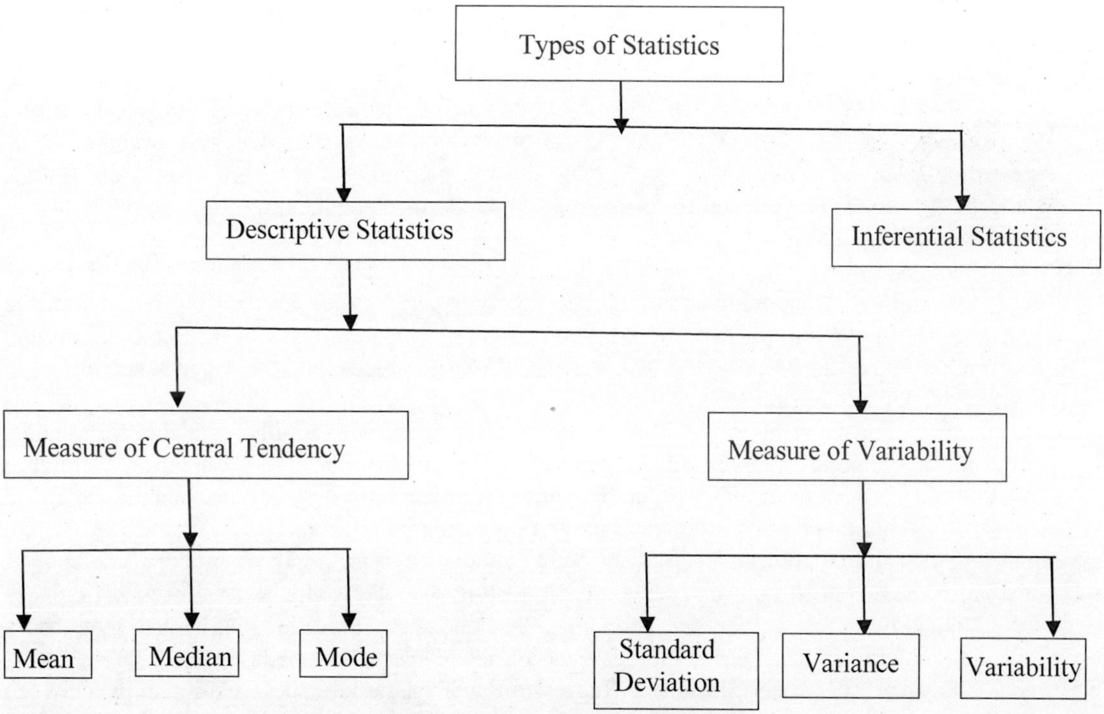

Fig. 3.1: Classification of Statistics

3.1.2 Definition of some Statistical Terms

Sampling: Sampling involves selecting a subset from the larger population to study and draw conclusions from.

Population: The population refers to the entire collection of items, individuals, or entities under consideration for data collection. It encompasses all the observations relevant to the researcher's study.

Sample: A sample is a smaller representative portion of the population chosen for analysis to make inferences about the entire population.

Census survey: A census survey entails the comprehensive examination of the whole population, encompassing a complete count of all its members.

Parameter: A parameter is a descriptive characteristic (value) derived from the entire population. It serves as a measurement of the population's attributes, such as the population mean and standard deviation.

Statistic (singular): A statistic is a measure employed to describe the characteristics of a sample. It is a value computed from the sample data and provides insights into the sample's attributes.

Sampling frame: The sampling frame is a compilation of individuals, items, or units from which the sample is selected.

Data: Data constitutes a collection of interconnected facts and figures that can be utilized to reach conclusions.

Variable: A variable denotes a specific attribute that varies across different objects and instances.

Sample size: The sample size corresponds to the count of elements or observations included within the sample

Data Collection and Descriptive Statistics:

Data collection involves gathering information on various parameters such as load, stress, temperature, and material properties. Descriptive statistics summarize and organize data, providing insights into central tendencies (mean, median, mode) and dispersion (range, standard deviation) of collected information. This aids in understanding the characteristics of materials and structures.

Probability and Probability Distributions: Probability theory quantifies the likelihood of events occurring and underpins many civil engineering analyses. Probability distributions model the distribution of potential outcomes in situations involving uncertainty. For instance, the normal distribution is used to describe the distribution of loads, stresses, and other variables, aiding in risk assessment and design.

Sampling and Sampling Distributions: Civil engineers often work with limited data due to practical constraints. Sampling involves selecting a subset of data from a larger population. Sampling distributions help engineers understand the variability of sample statistics, such as the mean and standard deviation, and make inferences about population parameters.

Inferential Statistics and Hypothesis Testing: Inferential statistics allow engineers to conclude about a population based on sample data. Hypothesis testing evaluates whether observed differences are statistically significant or due to chance. For instance, engineers might test hypotheses related to material strength or the effectiveness of new construction methods.

Regression Analysis and Model Building: Regression analysis establishes relationships between variables, enabling engineers to predict one variable based on others. In civil engineering, regression models can predict factors like structural deformation, traffic flow, or energy consumption based on relevant parameters. This aids in design and optimization.

Reliability and Risk Analysis: Reliability analysis assesses the probability of structures or systems performing satisfactorily under various conditions. Probabilistic methods help engineers design structures with desired levels of safety, considering factors like material strength, loads, and environmental conditions.

Quality Control and Statistical Process Control (SPC): In construction and manufacturing, quality control is essential to ensure that products and processes meet specified standards. Statistical Process Control (SPC) uses control charts to monitor the stability and variability of processes. SPC helps detect deviations early, preventing defects and ensuring consistent quality.

Geotechnical and Environmental Applications: In geotechnical engineering, statistical methods analyse soil properties, foundation behaviour, and slope stability. Environmental assessments involve analysing pollutant concentrations, air quality, and noise levels using statistical techniques.

Structural Health Monitoring (SHM): Structural Health Monitoring involves using sensors to monitor the condition of infrastructure in real-time. Statistical analysis of sensor data can detect anomalies, assess structural integrity, and predict maintenance needs.

Construction Management and Project Control: Statistics assist in project management by forecasting costs, durations, and resource requirements. Monte Carlo simulations model project uncertainties, helping managers make informed decisions to keep projects on schedule and within budget.

Spatial Data Analysis and Geographic Information Systems (GIS): Spatial data analysis uses statistical techniques to analyse data with geographic attributes. GIS integrates geographic data, enabling engineers to analyse and visualize information related to land use, infrastructure planning, and transportation networks.

3.2 DATA REDUCTION AND INTERPRETATION

Normally sensors are placed to collect the data either in raw form or in filtered form, consider the environmental effect, transfer the data to another sensor or to a data recording system and finally a data output system. Thus, each sensor handles a huge amount of data. However, many elements in the data set are redundant as well as irrelevant to the classification or identification process. Thus, an appropriate technique to reduce the data is very important. Before any discussion on data reduction, acquisition of data and its processing are to be understood very clearly.

Data acquisition consists of three steps:

a) Acquiring data from the measurand
b) Converting the acquired data to usable form
c) Displaying the usable data

3.2.1 Data acquisition

Some raw data like pressure, force, displacement, etc. are required to be acquired in analog form whereas some data like exceedance of a threshold, on or off of a motor, alarm to be activated or not are required to be acquired in digital form. When microprocessors are used in the data acquisition system, the data are handled digitally. A digital-to-analog (D/A) converter is then required to get analog information, whereas an analog to digital (A/D) converter is required to process the data. Again, for an A/D converter the sampling rate is also very important. Consider a sine wave of 1s period. If the sampling rate is 1 sec, the input data will be in phase with the sine wave and we will feel that we are measuring some non-varying input. If the sampling rate is 0.5 sec, there may be different output results depending on the chosen phase difference. This is called aliasing. If the sampling rate is 0.1 s, we expect more realistic data. Fig.3.2 shows the effect of the sampling rate on 1s sine wave.

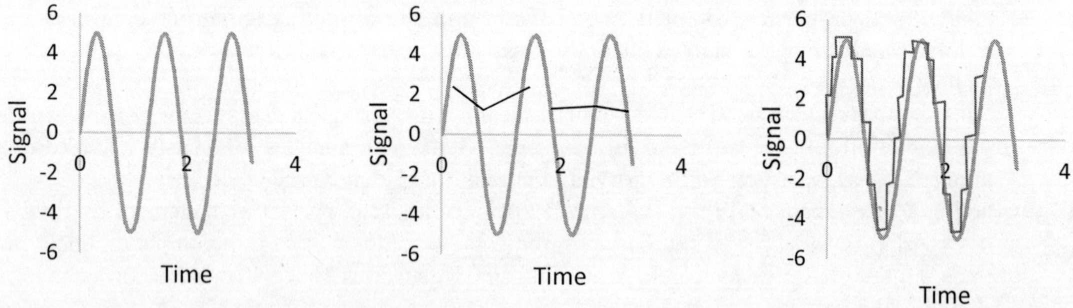

(a) True waveform (b) aliasing of data (c) sampling after discrete time

Fig. 3.2: Sine wave with 0.1 s sampling rate

Normally 8, 10, 12 bits D/A and A/D converters are used. The precision of these converters is shown in table 3.1 for a 6-volt measurement.

Table 3.1 Relation between precision and different bits of converter for a 6V measurement

Number of bits	Steps	Resolution on 6V measurement, V	% resolution
8	256	0.0234	0.390
10	1024	0.00586	0.098
12	4096	0.001465	0.0244

The accuracy and precision of acquired data often lead to confusion. Both of them are dependent on sampling interval and resolution. For the A/D conversion process, the step size or number of bits is critical for system accuracy and precision. The difference between accuracy and precision is illustrated in Fig 3.3

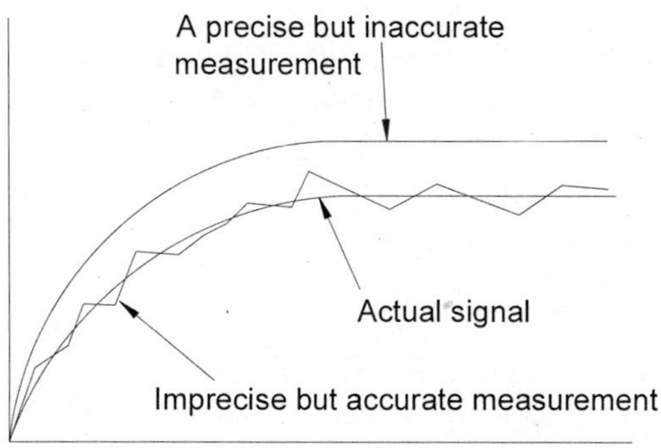

Fig. 3.3: Difference between accuracy and precision

3.2.2 Data Reduction

In many cases, the dimension M of a dataset may be very large and many elements are redundant or not required for the process. It is reported that Bayes' rule although powerful in handling any type of uncertainty cannot handle the complexity of a dataset. In general, most of the data reduction techniques are based on the selection of features. By features, we mean those components of the data set that are directly related to the inference problem. For example, if we monitor the damage to a building due to an explosion, then feature means those components of the data set which are directly related to the explosion. There are two ways to reduce a dataset, namely, feature extraction and feature selection. Feature extraction means the transformation of raw measurement space into the feature space. Feature selection means selecting the elements that are useful for measurement and discarding those elements which are irrelevant for the measurement.

As understood, feature extraction and feature selection from raw data require some basic knowledge of Artificial Intelligence (AI) (Kantardzic, 2003). Again, from AI consideration, the data reduction may be supervised or unsupervised. Supervised data reduction means transformation from data space to feature space is undertaken under some specific knowledge of the physical nature of the problem. In unsupervised cases, the transformation is data-driven. Mathematically, the transformation may be linear or non-linear. Since unsupervised techniques are more relevant, hence this technique is further discussed through following paragraphs.

Feature selection in general can be viewed as a search problem, specifying whether in a selected set the particular feature exists or not. For example, suppose a data set has three features [a1, a2, a3], and the presence of a feature is coded with 1 and absence with 0, then we have $2^3 = 8$ combinations of output selection sets like : [0, 0, 0], [1, 0, 0], [0, 1, 0], [0, 0, 1], [1, 1, 0], [0, 1, 1], [1, 0, 1] and [1, 1, 1]. The problem of feature selection becomes complicated in case the number of features increases, for N number of features, there will be 2^N number of sets and it becomes difficult to proceed with the above procedure. For practical reasons, an optimal search is not feasible, and simplifications are required to produce reasonable, acceptable and timely results. One possible technique for feature selection is based on the comparison of means and variances.

In general, for every feature, there may be a different number of classes. If only two classes A and B are considered and there are n1 samples in class A and n2 samples in class B, the following steps are adopted to ensure whether the feature can be adopted or not

a) Obtain mean and variance of the class A and class B
b) Obtain SE from the following equation:

$$\text{SE (A-B)} = \sqrt{\frac{Variance\ A}{n1} + \frac{Variance\ B}{n2}} \qquad (3.1)$$

c) Normalize the difference of mean with respect to SE and compare the normalized with the prescribed threshold value. If the normalized value is less than the threshold value the data can be reduced, otherwise not

Obtain $|(mean\ A - mean\ B)/SE(A-B)|$

Check the value with threshold value

The mean of a feature is compared in both classes without taking into consideration the relationship to other features. In this approach it is assumed a priori that each feature is independent of the others. Comparison of means is considered to be a natural fit for classification problems. For M classes, M pair-wise comparisons can be made between each class and its complement. The feature can be retained if the comparison does not exceed the threshold value. This can be explained through a problem.

Example 3.1

Ex. A dataset consists of two features A and B as shown below. The output feature C classifies each feature into two classes. If the threshold value is 0.5, which input feature can be accepted for data reduction?

Feature A	Feature B	Feature C
0.4	0.6	X
0.4	0.8	Y
0.5	0.7	X
0.6	0.6	X
0.9	0.9	Y
0.3	0.6	Y

Ans:

Data set for feature A and class X = [0.3, 0.6, 0.5]

Mean A_X = 0.5 and variance var(A_X) = 0.01

Data set for feature A and class Y= [0.4, 0.9, 0.3]

Mean A_Y = 0.5333 and variance var(A_Y) = 0.1033

Data set for feature B and class X= [0.4, 0.9, 0.3]

Mean B_X = 0.6333 and variance var(B_X) = 0.0033

Data set for feature B and class Y= [0.4, 0.9, 0.3]

Mean B_Y = 0.7667 and variance var(B_Y) = 0.0233

$$SE\ (A_X\text{-}A_Y) = \sqrt{\frac{Var(Ax)}{n1}+\frac{Var(Ay)}{n2}} = \sqrt{\frac{0.01}{3}+\frac{0.1033}{3}} = 0.1943$$

$$SE\ (B_X\text{-}B_Y) = \sqrt{\frac{0.0033}{3}+\frac{0.0233}{3}} = 0.0942$$

$$|(meanA_X - mean\ A_Y)/SE(A_X - A_Y)| = \frac{|.5-0.5333|}{0.1943} = 0.171 < 0.5$$

$$|(meanB_X - mean\ B_Y)/SE(B_X - B_Y)| = \frac{|.6333-0.7667|}{0.0942} = 1.416 > 0.5$$

Thus feature A can be reduced.

Example 3.1

For a large number of data, the above technique of data reduction may not be feasible. In such a case, another technique called Principal Component Analysis (PCA) may be used. Assume we have a dataset **C** with M number of sensors (s_1 ,s_2 , s_3, …, s_M), taking measurements for Nt time instances (t_1, t_2, t_3, …, t_{Nt}). Suppose $^{(i)}C_k$ is the reading of kth sensor s_k at time t_k. Thus, at kth time the dataset will be $\mathbf{C_k}$ which is a vector of size **Mx1**. The objective of the PCA is to extract an M-dimensional linear projection from M-dimensional data set. Generally, data are correlated and the degree of correlation is represented by the covariance matrix **COV.**

$$\boldsymbol{COV} = \frac{1}{N_t}\sum_{k=1}^{N_t}(\boldsymbol{C_k} - \mu)(\boldsymbol{C_k} - \mu)^t \tag{3.2}$$

Where μ is the mean value vector and is given by

$$\mu = \frac{1}{N_t}\sum_{k=1}^{N_t}\boldsymbol{C_k} \tag{3.3}$$

Let us define a new variable **x** to represent the coordinates of data features or Principal Components (PCs) given by

$$(\boldsymbol{C_k} - \mu) = \boldsymbol{A}\,\boldsymbol{x_k} \tag{3.4}$$

A is called the linear orthogonal transformation matrix and is obtained by solving the classical eigenvalue problem.

$$\boldsymbol{COV\ A} = \boldsymbol{A}\,\boldsymbol{\Lambda}(3.1) \tag{3.5}$$

Where, A is the diagonal matrix whose i[th] diagonal element is the ith eigenvalue λ_i of the covariance matrix ***COV.*** It is clear from equation 3.5 that **A** collects the eigenvector $\mathbf{a}_i$ of covariance matrix ***COV.*** The eigenvalues can be sorted as

$$\lambda_1 \geq \lambda_2 \geq \lambda_3 \geq \ldots \geq \lambda_n \geq 0$$

Since λ represents the variance of each PC of the dataset, the higher the value of λ, the better will be the information carried over. To reduce the initial M-dimensional dataset to an m-dimensional set, we can represent it by the best subset of m-PCs. The most straightforward criterion is to select a subset of components whose cumulative variance is greater than the desired percentage $\bar{R}$ (say, 90%) given by

$$R_m = \frac{\sum_{i=1}^{m} \lambda_i}{\sum_{i=1}^{M} \lambda_i} = \frac{\sum_{i=1}^{m} \lambda_i}{tr(COV)} \geq \bar{R} \quad (3.6)$$

Where, tr(***COV***) is the trace of the covariance matrix

3.2.3 Data interpretation

Once data collected, they must be properly interpreted to draw correct conclusions and use the information. Data interpretation is the process of reviewing data and arriving at some relevant conclusions using various analytical research methods. There may be two types of data, calibrated data and non-calibrated data. Calibrated data means the sensor is already calibrated and we are getting the direct outputs; like a pressure sensor is giving data 50 kPa, 100 kPa, … etc. The non-calibrated data means the sensor output is not calibrated and we are to train the outputs to our desired requirement. Let us discuss the interpretation of both types of data.

Calibrated data:

There may be three types of data:

(a) Quantitative
(b) Qualitative
(c) Mixed method

In quantitative data interpretation, the numerical data are analysed. Thus, it will directly meet our requirements. We can easily find how much data is available within a range, or what is the maximum value or minimum value, etc.

In qualitative data interpretation, non-numerical data like text, audio, image, etc. are to be analysed. This can be used to identify the trends and patterns.

In the mixed method, both numerical and non-numerical data are interpreted.

Again, there are three methods by which data can be interpreted :

(a) Descriptive statistics

This method includes the application of simple statistics like finding the mean, mode, standard deviation, etc to represent the data in understandably.

(b) Inferential statistics

This method includes standard statistical models and algorithms to identify patterns and relationships in the data.

(c) Visualization technique

This method involves creating visual representations like graphs, charts, maps to represent very complex data in an easily understandable form.

Uncalibrated data:

For uncalibrated data, the input data are to be inducted first, it is to be learned for a desired output. The learning method is an algorithm, usually incorporated in a software which maps the inputs and outputs. Once mapping is completed, the inputs will be directly used to obtain the outputs. This inductive learning also called machine learning is a combination of artificial intelligence and statistics. In the theory of inductive learning, all input data are organized and each pair of input and corresponding output is called a sample. The general learning scenario consists of three components as shown in Fig 3.4.

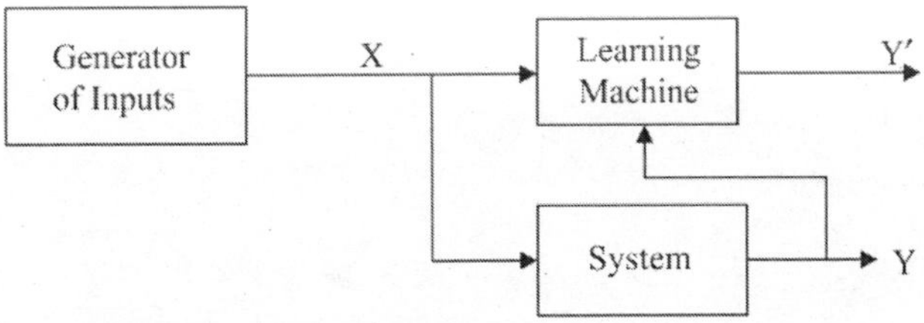

Fig. 3.4: Three components of learning process

The components are

(a) Generator, to generate input vector X
(b) A system that generates an output vector Y
(c) A learning machine that estimates an unknown mapping between input X and output Y' from the observed sample X and Y.

To explain the learning process, suppose there is a collection of samples X and Y. The unknown mapping yields a function Y'. The function Y' is often called a hypothesis. Fig 3.5 shows a simple example of applying the hypothesis. Fig 3.5 (a) shows a set of values X and Y. One hypothesis may be a straight line running with a minimum distance from the data points Fig 3.5 (b). The others may be a polynomial (Fig 3.5 (c)) or a parabolic (Fig 3.5 (d)). It is a very difficult task to select the best possible hypothesis. Hence, some a priori knowledge may be required to select the most appropriate hypothesis.

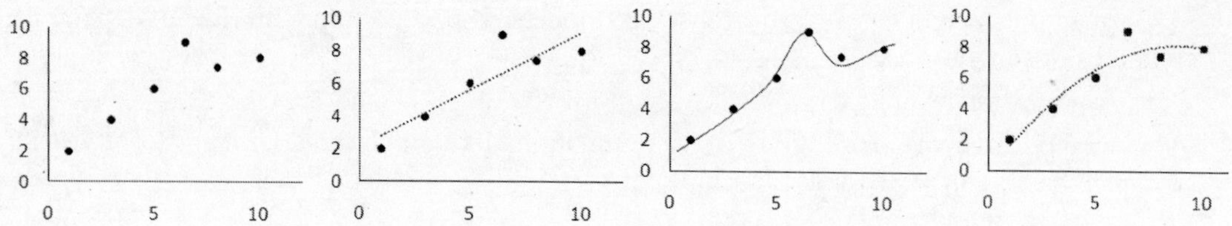

(a) (b) (c) (d)

Fig. 3.5: Three hypotheses for a given data set

Thus, for a set of uncalibrated data, following are the different techniques of interpretation:

1. Classification
2. Regression
3. Clustering
4. Summarization
5. Dependency modelling

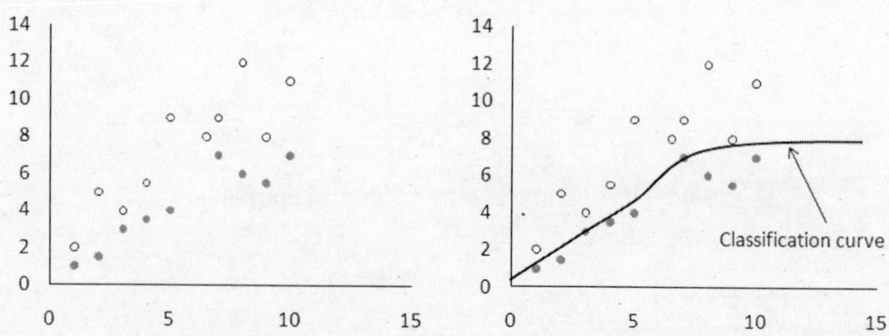

(a) Training data set (b) Classification curve

Fig. 3.6: Graphical interpretation of classification

The initial training data may contain different classes. Different graphical symbols are used to visualize each class as shown in Fig 3.6. The final result of classification in a 2-D space is the curve shown in Fig. 3.6 (b). This curve separates the classes. Support Vector Machine (SVM) is a successful tool to solve the classification problem. It consists of developing a function called hyperplane to separate the classes of samples. The hyperplane is a line when the number of input features is two and a plane when the number of input features is three as shown in fig 3.7. It is difficult to visualize the hyperplane if the number of input features is more than 3.

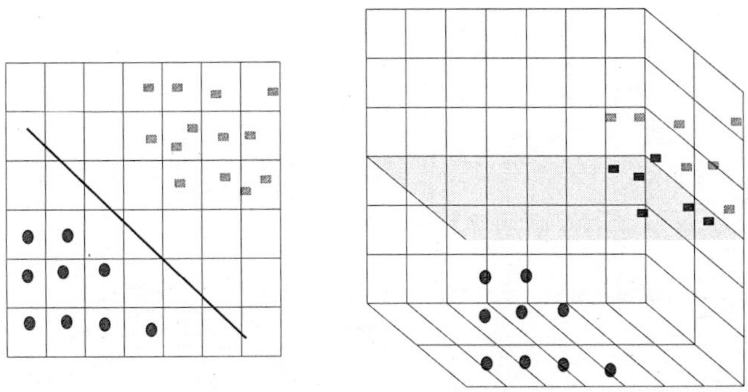

(a) Hyperplane is a straight line (b) Hyperplane is a plane

Fig. 3.7: Hyperplane in 2D and 3D space

Consider an example shown in fig 3.8 where the dark spots and red spots are to be separated. There may be a large number of hyperplanes fig 3.8(a). How one can select the best one – that is the main question. The main idea is to locate the decision boundary at the maximum distance from the closer data points called the support vectors. To achieve this we have to plot a large margin. A margin is defined as the distance or separation from the two hyperplanes connecting two classes. A narrow margin and a large margin are explained in Fig 3.8. Thus the optimal SVM boundary will have the maximum margin. For the maximum margin, the equation of the boundary line is obtained. For any pair of input data if the point lies above the line, the value of f(x) will be > 0. An example is shown below.

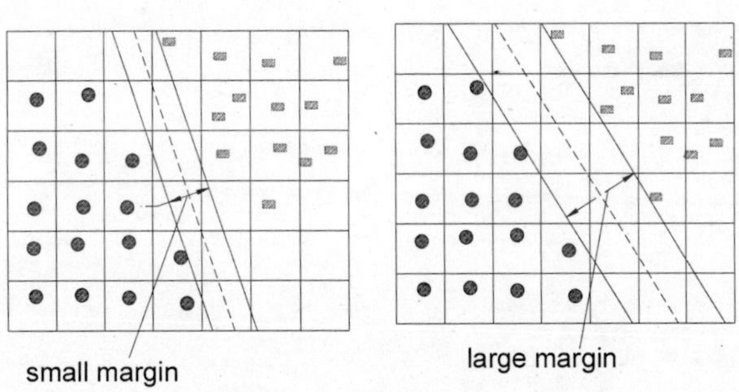

Fig. 3.8: SVM boundary

Ex. In an SVM model the boundary equation between two classes is obtained to be f(x) = x1 + 2 x2 -5 =0 as shown below. Obtain which class the point (2,1) will belong to ?

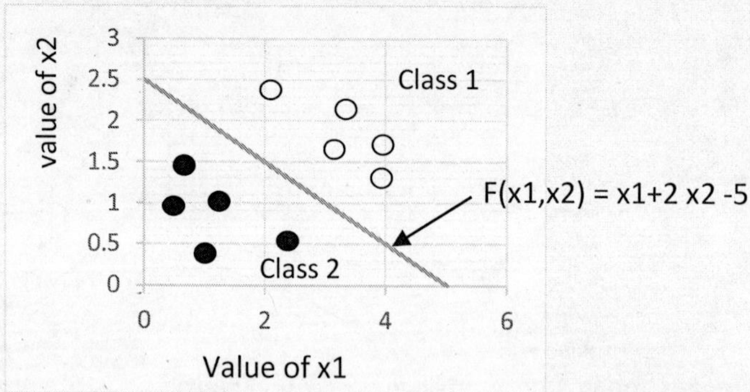

Ans:

Given, f(x1, x2) = x1 +2 x2 -5 = 0

For the input data (2,1), f(x1, x2) = -2 which is less than 0

Hence the data will fall in class 2

Example 3.2

In regression, a function is obtained from the input training data. For the input training data as shown in fig 3.9 (a). A regression function is obtained as shown in fig 3.9 (b). Based on this function, it is possible to estimate the value of a prediction variable for each new sample.

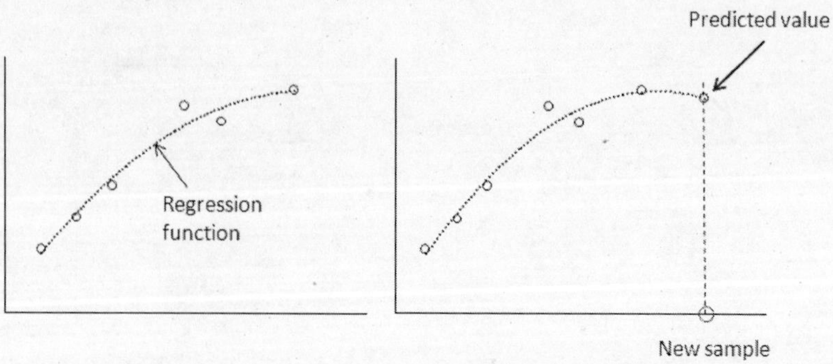

a) Training data set (b) Regression function

Fig. 3.9: Graphical interpretation of regression

Clustering is the most common unsupervised learning task. Fig 3.10 (a) shows the input data. Fig 3.10 (b) shows the different groups or clusters to describe the data.

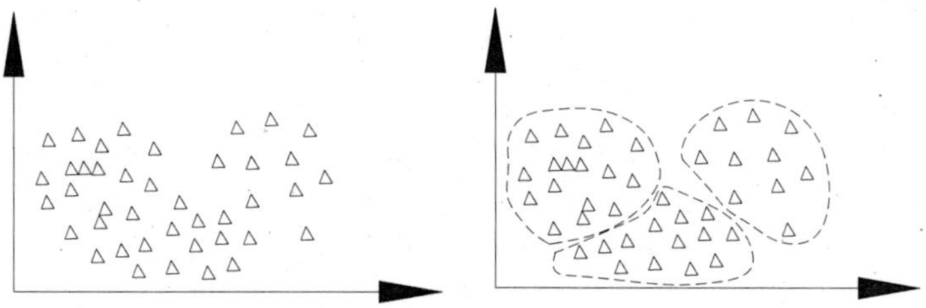

a) Training data set (b) Clustering

Fig. 3.10: Graphical interpretation of clustering

Summarization is a process where the limits are set to the input data. Fig 3.11 (a) shows the input training data and Fig 3.11 (b) shows the process of summarization.

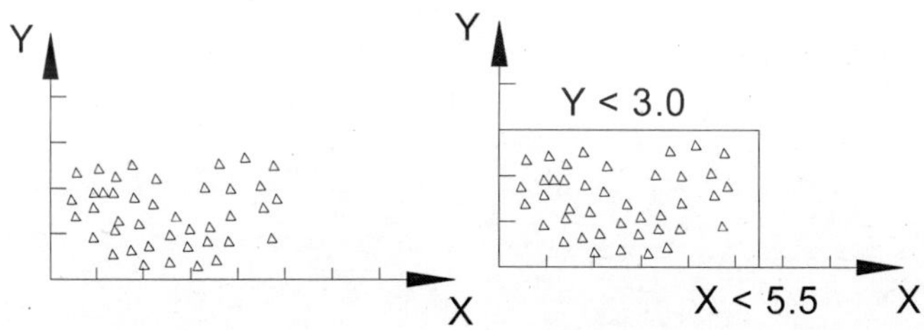

a) Training data set (b) Boundary demarcation

Fig. 3.11: Graphical interpretation of summarization

Whatever the input data and the probable data mining techniques are, it is to be understood that the success of a proper data mining depends upon the data-pre-processing and data-reduction methods. When the data are pre-processed and we know what kind of learning task is defined for our application, there are many data-mining and knowledge discovery techniques as follows :

1. Statistical methods: Bayesian inference, logistic regression, analysis of variance(ANOVA), etc.
2. Cluster analysis : Divisible algorithms, agglomerative algorithms, partitional clustering, incremental clustering
3. Decision trees and decision rules : Concept learning system (CLS), ID3 algorithm
4. Artificial Neural Networks
5. Genetic algorithms
6. Fuzzy inference systems

3.3 DATA COLLECTION AND GENERAL OPERATIONS OF SOME INSTRUMENTS

A small description of some instruments are narrated below to show how they are used in different applications and how data are collected from them.

3.3.1 Piezometer

Vibrating wire (VW) type piezometer is discussed here. The piezometer consists of a tensioned steel wire which is attached to a flexible diaphragm at one end and to the inner body. The diaphragm is the sensing element. The entire assembly is enclosed in a steel body. Two opposing coils are located within the steel casing near to the wire. When a voltage is applied to the coils, a magnetic field is created. The wire vibrates at the resonant frequency in the magnetic field and generates an alternating current. The frequency of this alternating current is detected and processed by the data-logger or vibrating wire read out unit which gives the output in terms of pressure unit. When a pressure is applied to the diaphragm, a tension is developed in the wire resulting in a change in the resonant frequency. The square if this frequency of oscillation is proportional to the applied pressure. Fig 3.12 shows a vibrating wire type piezometer. Picture of a vibrating wire piezometer is shown in Fig 3.13

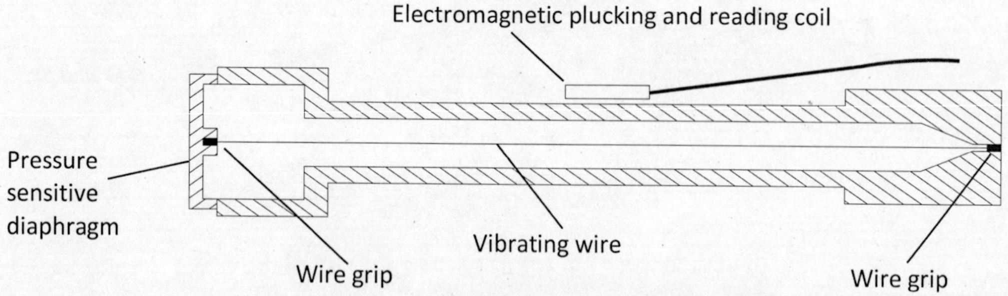

Fig. 3.12: Vibrating wire type piezometer

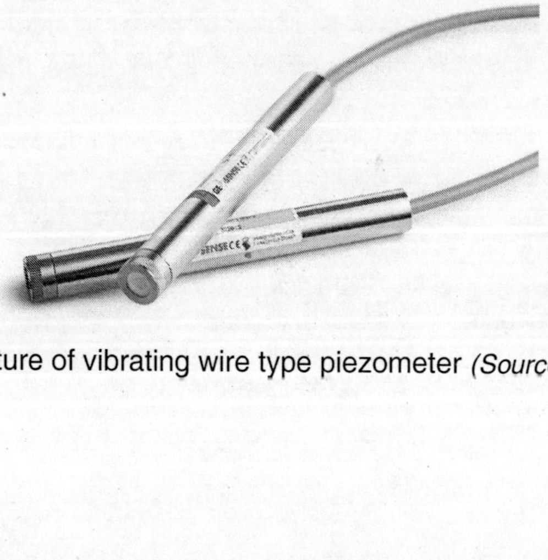

Fig. 3.13: Picture of vibrating wire type piezometer *(Source : Geoscience)*

Installation

The piezometer can be installed in cased or uncased boreholes. In cased borehole, one casing is to be inserted after the borehole is constructed. If the bottom or the borehole is to be plugged, pallets of bentonite is to be poured once the depth of installation of the piezometer is reached. The piezometer is lowered. Filter sand is then poured so that the piezometer is placed at the required location. More clean filter sand is put on the top of the piezometer to provide the required filter above the piezometer level. If required bentonite seal of specified thickness may be made on the top of the filter sand. Fig 3.14 shows installation of a piezometer in ground

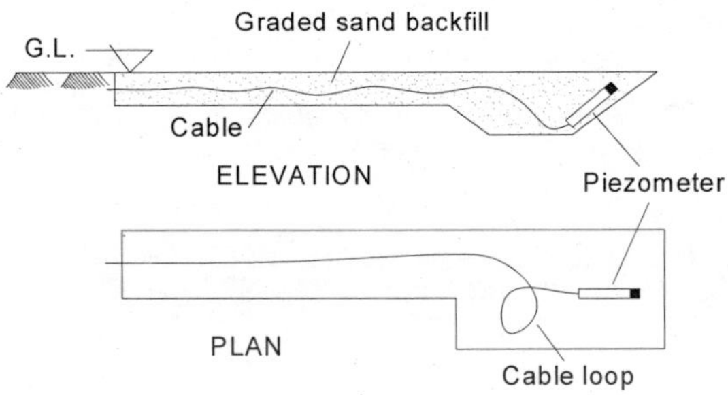

Fig. 3.14: Typical example on installation of a piezometer in ground

Piezometer readings

Most of the piezometers contain pressure and temperature sensors. The temperature sensor depicts the temperature of the ground water which has an effect on the water pressure. The read out unit displays the reading in terms of $Hz^2/1000$ or $Hz^2/10000000$. Since the reading is obtained in terms of frequency, hence the data is to be reduced to pressure unit. The data reduction is obtained by applying calibration factors. For an instrument, this calibration factor is unique and is supplied by the manufacturer.

Initially the reading of piezometer without any water pressure is to be taken. This is called base pressure or site zero reading. The procedure for site zero reading is as follows:

Take a bucket and fill it with clean water. The temperature of water should preferably be equal to the ground water temperature. The water is to be stirred for a while so that the temperature is uniform across the depth. The piezometer is taken and is immersed in the bucket water. If there is any rubber seal at the tip of the piezometer, the rubber seal is to be removed. The filter cap from the tip of the piezometer is also to be removed. All these operations are to be carried out keeping the piezometer immersed into the water. Switch on the readout unit. When the reading is stable hold the wire and lift the piezometer out of the bucket allowing it to hang vertically downwards and immediately record 2 or 3 readings. Replace the transducer to the water, wait for 2 to 3 minutes. The piezometer is taken out of the bucket and another set of readings are taken. Repeat at least three times and take the average reading which is termed as base pressure or site zero reading.

The piezometer is then lowered and installed in the borehole at the required depth. The reading of the piezometer is taken. The readings will be in Hz. The site zero reading and the present site reading are then converted to linear digits :

$$\text{Linear digits} = \frac{1}{1000 * (Hz * 10^{-7})^2} \tag{3.7}$$

The water pressure in terms of water head in m can be obtained as follows:

Water pressure, m = K * ((current reading – base reading) in linear digits)

Where K is the calibration factor.

A sample calculation of the piezometer reading is shown below.

Ex. Obtain the water pressure from a piezometer with the following readings:

Site zero reading = 3255

Current reading = 3278

Calibration factor for water pressure in m = -0.01820

Zero reading converted to linear digits = $(1 / (3255\text{x } 10^{-7}))^2/1000$

= 9438.4

current reading converted to linear digits = $(1 / (3278\text{x } 10^{-7}))^2/1000$

= 9306.4

Water pressure in m = K x (current reading – base reading)

= -0.01820 * (9306.4 – 9438.4) = 2.40 m

Example 3.3

The pore pressure can also be obtained from polynomial equation using the linear digit reading. Normally three factors are used, the calibration unit for all these three factors are supplied by the manufacturer. As an example, the above problem may be solved with polynomial equation as shown below.

Ex. Obtain the water pressure from a piezometer with the following readings:

Site zero reading = 3255

Current reading = 3278

Calibration factors for water pressure in kPa for factors A= -6.30×10^{-7}

B = - 0.1685

C = 1647

Ans.

Current reading in linear digits = 9306.4

Water pressure in kPa = A x $(\text{reading})^2$ + B x (reading) + C

$= -6.30 \times 10^{-7} \times (9306.4)^2 + (-0.1685) \times (9306.4) + 1647$

$= -54.56 - 1568.128 + 1647 = 24.31$ kPa

Example 3.4

3.3.2 Inclinometer

As the name implies, the inclinometer measures the angle of inclination of an object with respect to some reference plane. These instruments have a wide application, namely, mobile phones, tablets, automotive and aerospace systems, in construction and mining equipment, etc. the sensor consists of a mass suspended within a housing by means of a flexible support like spring or a flexure element. When the sensor is tilted, the mass is shifted from its position within the housing resulting in a change in the position of the mass relative to the housing.

The change in position is detected by a sensor which may be of different types like capacitive, inductive or piezoelectric, etc. The sensor measures the displacement of the mass. The output of these sensors is then processed by a microcontroller or other electronic circuit to calculate the angle of inclination based on the measured displacement.

There are two types of accelerometers which are used in the inclinometer probes. They are servo-accelerometer and MEMS accelerometer.

- **Servo-accelerometer**: The servo-accelerometer probes are available in the market since past. In an accelerometer, the relative position of two electrodes – one fixed to the instrument's body and the other suspended in a way that allows it to move freely along one axis – produce a capacitance. When the instrument is level – either vertically or horizontally, the position of capacitance is considered to be level. When the instrument tilts, the floating electrode moves along its axis. There is a change in capacitance which gives the instrument orientation. The force balanced sensing element measures the tilt from the vertical of the probe. The probe contains two sets of spring pressured wheels which run along the longitudinal grooves of the guide casing. A maximum system precision of 1.2 mm per 30 m is possible with this instrument. The resolution is nearly linear and constant for an inclination 0 to $\pm 30^0$ from vertical.

- **MEMS accelerometer :** The MEMS (micro electro mechanical system) technology is recently introduced. Here MEMS type sensor is used instead of servo-accelerometer. There are many advantages of this type of accelerometer, like they are cheap, low power consumption, durability, wireless transmission facility, etc.

The manual inclinometer consists of following parts :

1. Inclinometer probe
2. Inclinometer reel Readout unit
3. Accessories like battery and charger.

Different parts of the inclinometer are shown in Fig. 3.15:

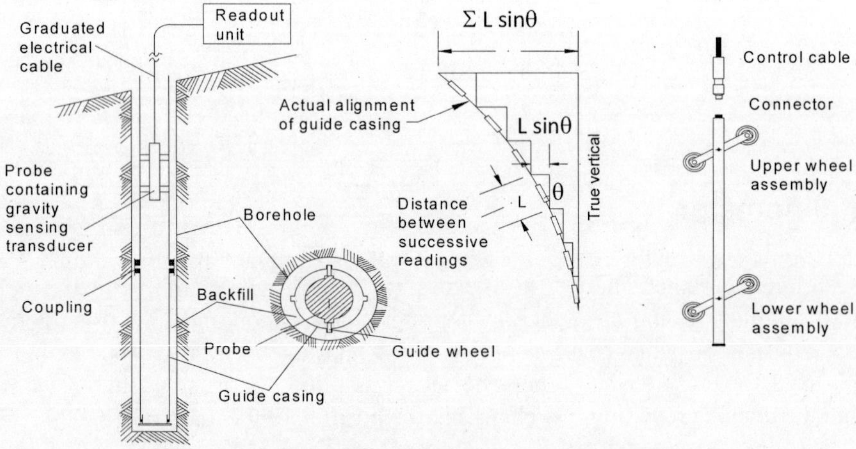

Fig. 3.15: Different parts of an inclinometer

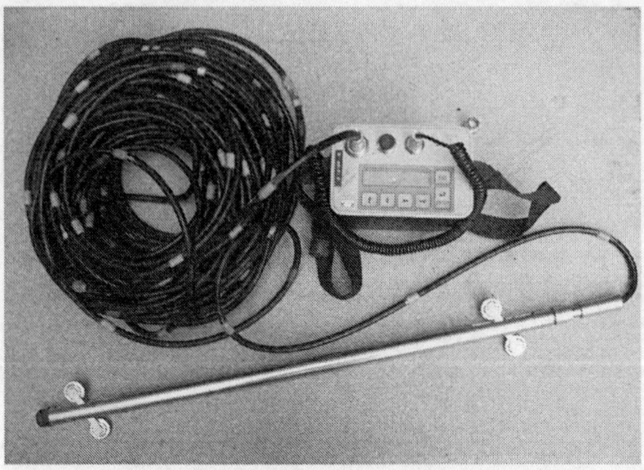

Fig. 3.16: Photograph of an inclinometer

(source : Transportation research circular E-C 129)

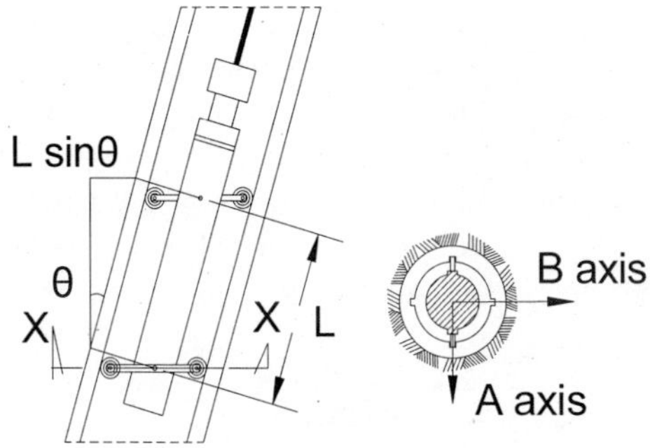

Sec X-X

Fig. 3.17: Probe with an inclination θ

Fig 3.17 shows the probe in the casing and inclination, θ of the casing with respect to the vertical. The probe is connected to a power source and a read out unit to obtain the measurement. The measurement starts from the bottom and the probe is raised incrementally at an interval of 0.3 to 0.6 m till upto the top of the borehole. The measurement starts with zero reading, i.e. when the probe is in vertical position. The difference from the vertical is the change in the shape and position of the probe. If there is a landslide, the casing moves along the direction of the landslide

Acceleration based inclinometer uses changes in acceleration for measuring the angle. Initially the inclinometer is positioned with some reference axis so that the orientation of the gravity is interpreted with the sensor's axes. The accelerometers continuously measure the accelerations. When the device gets tilted the accelerometer measures the acceleration due to gravity along the vertical axis (z-axis). Using trigonometry the inclinometer measures the angle based on change in acceleration along the vertical axis. This angle is the required angle of tilt. The accelerometers can also be used for dynamic measurement also. The output is shown digitally on a screen either in degree or in radians

Example 3.5

Ex. In an acceleration type inclinometer, three accelerometers are placed in three perpendicular axes. The inclinometer is laid horizontally and the accelerometer along z axis measures an acceleration of 9.8 m/s^2. The inclinometer is tilted and the above accelerometer shows and acceleration of 9.0 m/s^2.

Ans. : when the accelerometer shows 9.8 m/s^2, the accelerometer is placed horizontally. When the acceleration is 9.0 m/s^2, the angle of tilt is given by :

θ= atan (9.0/9.8) = 42.56 ^{0}

Data acquisition in MEMS digital inclinometer system:

The digital inclinometer probe contains two MEMS accelerometers which measure tilt in two axes. One axis termed as 'A' axis is aligned to the expected movement direction. One set of grooves will be along the A axis. The other set of grooves is at perpendicular direction and termed as 'B' axis. Both the axes are shown in fig 3.16. Although it may be difficult to place the probe accurately along the movement, however, attempt is to be taken to place the probe along the possible direction of the movement. The upper wheels of the probe are placed along the slope direction. The groove closest to the anticipated movement direction is usually chosen as the main reference direction. This reference direction is marked as A+. The readings are taken from the bottom towards the top at the specified depth. The readings mainly consist of deviations from the vertical axis over the distance between the upper and lower wheels (Fig. 3.18). All measurements are taken along this reference axis.

$$D = L \sin\theta$$

D = lateral displacement in the direction of the axis of measurement

L = inclinometer probe length, usually 0.5 m

θ = inclination of the probe with respect to vertical axis

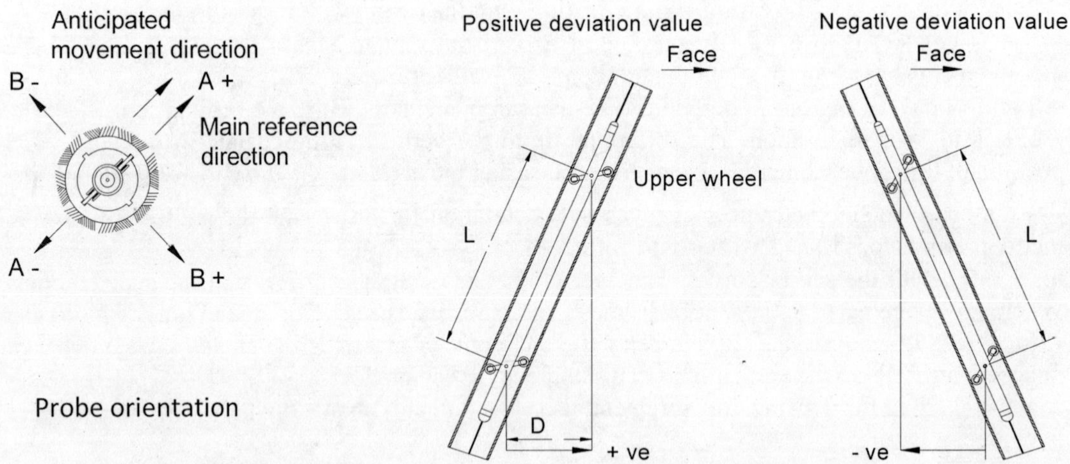

Fig. 3.18: Sign convention in the A axis and deviation D measured by the probe

Once the borehole is dug, it is to be washed property. The guide casing is inserted. Insert the probe into the borehole aligning the upper wheel in the direction of A+. The probe is then lowered to the desired depth slowly and smoothly. The reading is then taken. The probe is then pulled up to the next measurement depth and the reading is taken. The profiling is continued till the top of the borehole. The probe is then again inserted into the borehole with 180^0 turning so that the upper wheels are along A- groove and lower wheels are along A+ groove. The observations are repeated.

The other accelerometer measures the readings along B axis. Three types of graphs are plotted along the depth :

1. Mean deviation along depth
2. Absolute deviation along depth
3. Checksum along depth.

The values are obtained as follows :

Mean deviation :along A axis = (A+ readings – A- readings)/2

along B axis = (B+ readings – B- readings)/2

Absolute position : summation of mean deviation starting from bottom upto the desired depth.

Check sum : along A axis = A+ readings + A- readings

along B axis = B+ readings + B- readings

when two sets of readings are compared to obtain the tilting over a certain time, then cumulative displacement is to be found out. It is the difference between two mean deviations and gives actual displacement of the borehole over a given time.

Ex. Obtain the mean deviation, absolute position and checksum from the following inclinometer observations along A axis.

Depth , m	Face A+	Face A-	Depth , m	Face A+	Face A-
0.5	0.005	-0.0054	6	-0.0032	0.003
1	0.005	-0.0053	6.5	-0.0036	0.0033
1.5	0.0039	-0.0041	7	-0.0031	0.0028
2	0.0029	-0.0031	7.5	-0.0035	0.0032
2.5	0.002	-0.0023	8	-0.0055	0.0053
3	0.0022	-0.0024	8.5	-0.0065	0.006
3.5	0.0007	-0.0008	9	-0.0039	0.0036
4	0.0004	-0.0007	9.5	-0.0022	0.0019
4.5	-0.0001	-0.0001	10	-0.0009	0.0007
5	-0.0036	0.0032	10.5	-0.0009	0.0007
5.5	-0.006	0.0056			

Ans:

Depth , m	Mean deviation	Check sum	Depth , m	Mean deviation	Check sum
0.5	0.0052	-0.018	6	-0.0031	-0.0319
1	0.00515	-0.0232	6.5	-0.00345	-0.0288
1.5	0.004	-0.02835	7	-0.00295	-0.02535
2	0.003	-0.03235	7.5	-0.00335	-0.0224
2.5	0.00215	-0.03535	8	-0.0054	-0.01905
3	0.0023	-0.0375	8.5	-0.00625	-0.01365
3.5	0.00075	-0.0398	9	-0.00375	-0.0074
4	0.00055	-0.04055	9.5	-0.00205	-0.00365
4.5	0	-0.0411	10	-0.0008	-0.0016
5	-0.0034	-0.0411	10.5	-0.0008	-0.0008
5.5	-0.0058	-0.0377			

The results are shown in Fig 3.19

Example 3.6

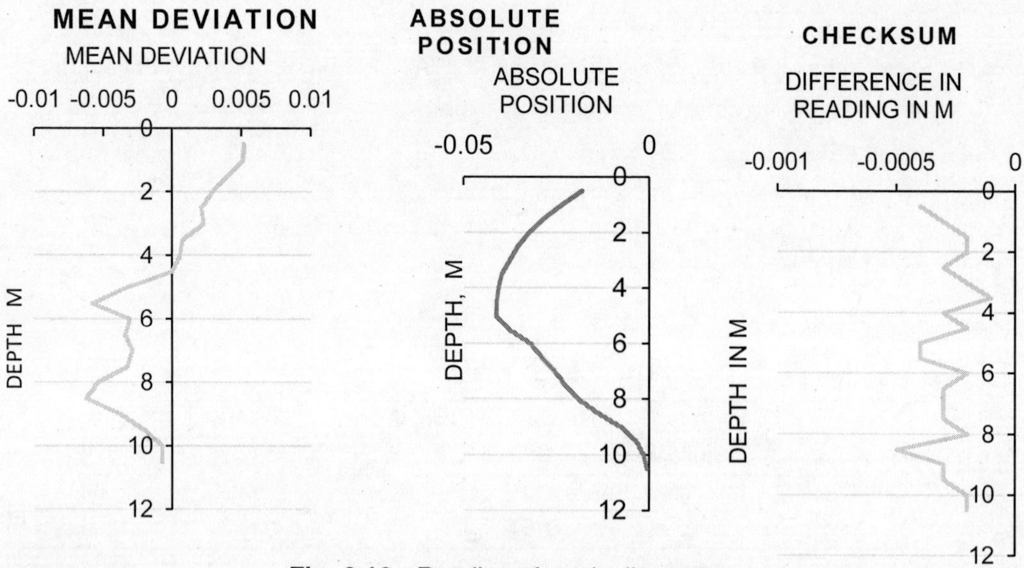

Fig. 3.19: Readings from inclinometer

3.3.3 Strain Gauge

Strain gauges are used to measure strain. Different components of a strain gauge are shown in Fig. 3.20. As the name implies, the strain gauge measures the strain. If tension is applied to the strain gauge, there will be increase in length. The increase in length will increase the resistance. The strain gauge is insensitive to the lateral forces. The working of a strain gauge is shown in Fig 3.21

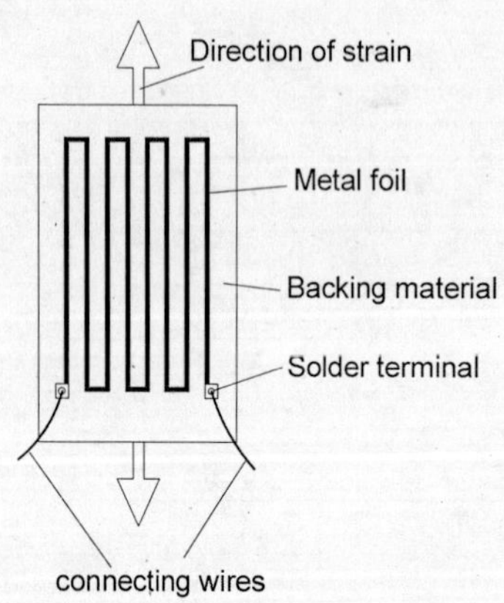

FIg. 3.20: Components of a strain gauge

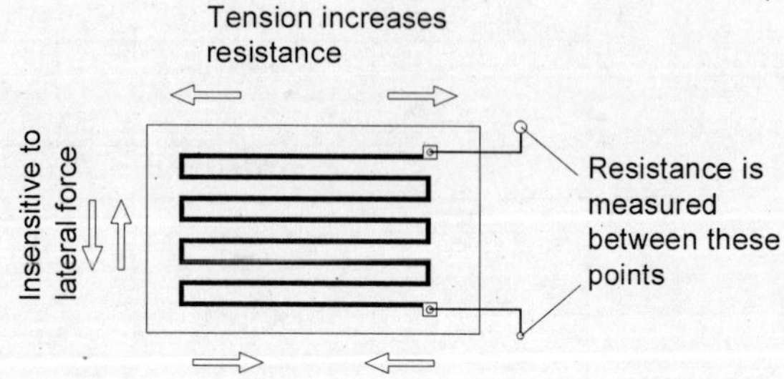

Fig. 3.21: Working of a strain gauge

The strain gauge has a gauge factor. Suppose a strain gauge has a gauge factor 5, it means that for every unit change in strain, the resistance of the strain gauge will be changed by 5 times its original value.

Let us take an example :

Suppose the original resistance of a strain gauge is 100 Ω which is under no load condition. Under a load let there be a strain of 0.001 and we are to find out the change in resistance. This can be easily found out as under :

Change in resistance :ΔR = gauge factor x strain x original resistance

$$= 5 \times 0.001 \times 100 = 0.5\ \Omega$$

Picture of a strain gauge is shown in Fig 3.22

Fig. 3.22: Picture of a strain gauge

Selection of a strain gauge:

a) Selection based on length : Strain gauges are available in different lengths, ranging from 0.2 mm to 100mm. However, for general applications the recommended length is 3 mm to 6 mm. Shorter range (< 3mm) is recommended if the space for installation is small, like a 25 mm diameter slot. The higher length is recommended if the space for installation is large, material is non-homogeneous like concrete and temperature is not an issue. Larger strain gauges are more expensive.
b) Selection based on gauge resistance : The higher the gauge resistance the higher is the sensitivity. A higher gauge resistance 350Ω to 1000Ω is selected when high sensitivity is required. For fatigue testing, a low resistance strain gauge 120 Ω is selected.
c) Selection based on gauge pattern: for measurement of a uniaxial strain, single strain gauge oriented along the direction of strain is used. A $0^0 - 90^0$ Tee rosette strain gauge is used when the principal stresses are to be found out. For three dimensional stress analysis, a 0^0- 45^0 - 90^0 rectangular rosette or 0^0 - 60^0 - 120^0 delta rosette is used. All the gauge patterns are shown in Fig 3.23

3.4 TIME DOMAIN SIGNAL PROCESSING

A signal is a function of an independent variable like time, distance, position, temperature, pressure etc. For example

a) Speech signal, image signal in electrical engineering
b) Vibration in civil and mechanical engineering
c) Radiation in physics
d) Space signals in astronomy
e) ECG, EEG, MRI, etc in Biotechnology
f) Earthquake vibrations in seismology, etc

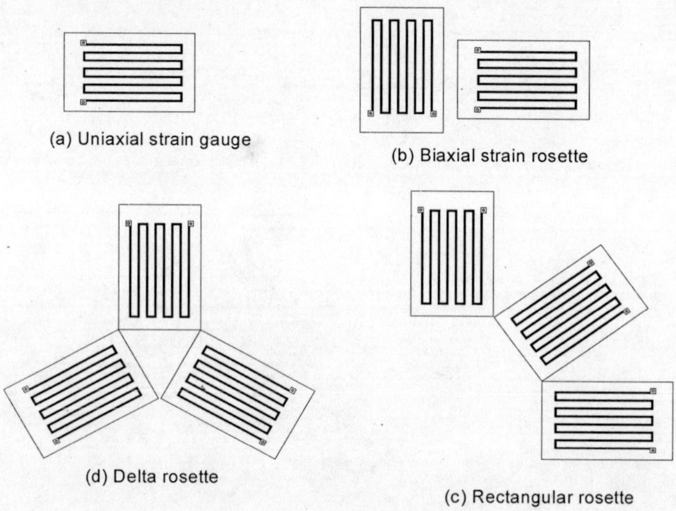

Fig. 3.23: Different gauge patterns

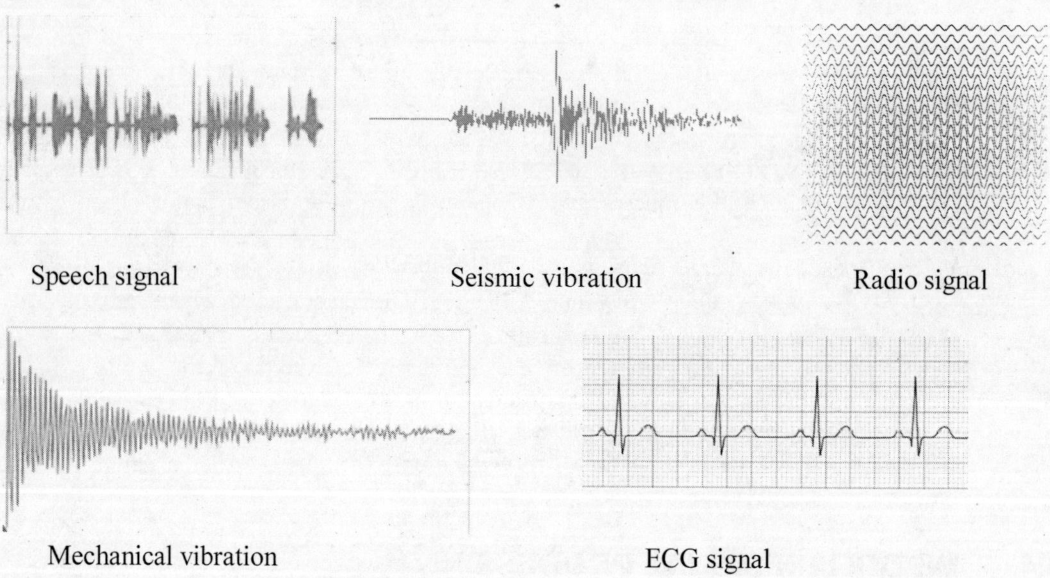

Fig. 3.24: Different types of time domain signals

A time domain analysis means an analysis of a signal with reference to time. With this definition, examples shown in Fig 3.24 are the time domain signals.

Signals in time domain may be continuous time signal or discrete time signal. Following are some characteristics of a continuous time signal:

1. It is a continuous time variable which has infinite number of values in its sequence.
2. It is also called analog signal and can be defined at any instant of time
3. It is represented on a real time axis. The amplitude is known at any instant of time
4. Examples of continuous signals are sine waves, cosine waves, triangular waves, etc. Any electrical signal associated with pressure, temperature, sound, etc are also continuous signals.
5. It is usually represented by x(t), where t is any instant of time

The discrete signal has the following characteristics:

1. It is a discrete time variable which has countable or finite set of numbers in its sequence. Normally a continuous system is represented digitally by a discrete system.
2. The output from a computer data is an example of discrete time signal.
3. The amplitudes are given at certain instances of time over the sequence.
4. The digital time signal is defined only at the sampling instants. The signals are represented with binary bits and stored in a digital medium.
5. It is usually represented by x(n), where n is any discrete time.
6. The amplitude between two successive time instants is not defined.

Difference between an analog signal and discrete time domain signal is shown in Fig 3.25

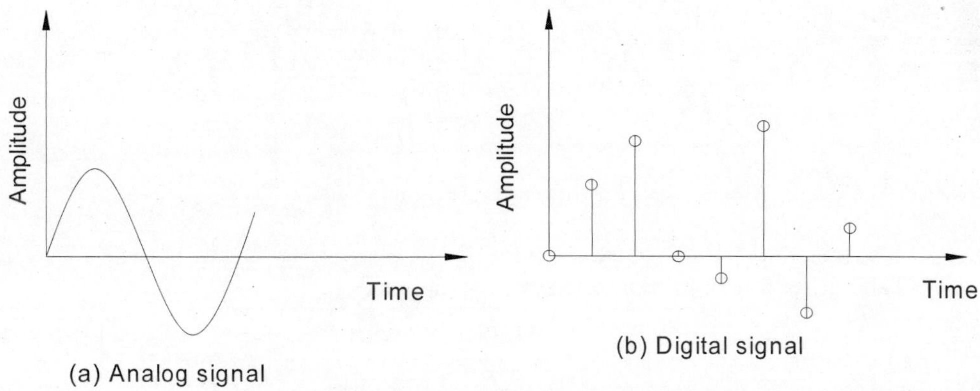

Fig. 3.25: Analog signal and digital signal

3.4.1 Classification of discrete time signal

The discrete time signal is represented by four ways:

a) Graphical
b) Functional
c) Tubular
d) Sequence

Discrete signal x(n) may have values like

X(-2) = 1; x(-1) = 3; x(0) = 0; x(1) = 4; x(2) = -2; x(3) = 5;

The signal is represented graphically as shown in Fig. 3.26 :

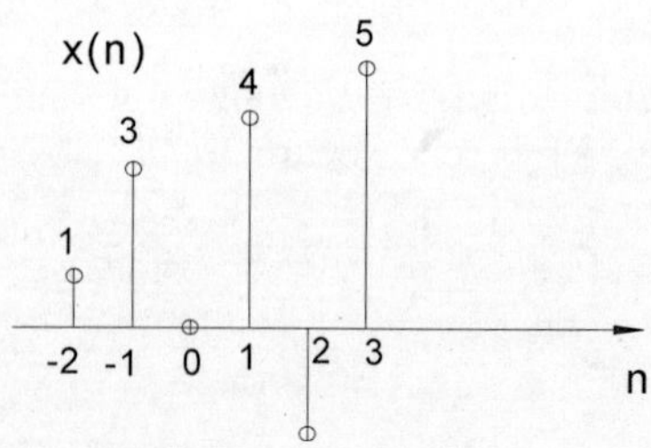

Fig. 3.26: Graphical representation of a digital signal

The above example may be represented as a function like

$$x\,(n) = \begin{Bmatrix} 1\ for\ n = -2 \\ 3\ for\ n = -1 \\ 0\ \ for\ n = 0 \\ 4\ \ for\ n = 1 \\ -2\ \ for\ n = 2 \\ 5\ \ for\ n = 3 \end{Bmatrix}$$

Tabular representation of the above example is :

n	-2	-1	0	1	2	3
x(n)	1	3	0	4	-2	5

Sequence representation of above example is:

$$\text{x(n)} = \left\{ \begin{matrix} 1 & 3 & 0 & 4 & -2 & 5 \\ & & \uparrow & & & \end{matrix} \right\}$$

the arrow mark ↑ denotes n = 0

3.4.2 Discrete signal

There are some elementary discrete signals which serve as basic building blocks for construction of more complex signals. These elementary signals are used to model a large number of physical signals, which occur in nature. The following are some elementary signals:

a) Unit step sequence
b) Unit ramp sequence
c) Unit parabolic sequence
d) Unit impulse sequence
e) Sinusoidal sequence
f) Real exponential sequence
g) Complex exponential sequence

Unit step sequence :

The step sequence is a basic signal for many complex signals. It exists for positive time and is zero for negative time. Its magnitude suddenly increases and remains constant at the sampling instants. In between the sampling instants its value is zero. If the magnitude is unity, it is called unit step sequence.

The discrete time unit step sequence u(n) is given by:

$$u(n) = 1, \text{ for } n \geq 0$$
$$= 0 \text{ for } n < 0$$

For shifted discrete time unit step sequence u(n -k) is given by :

$$u(n - k) = 1, \text{ for } n \geq k$$
$$= 0 \text{ for } n < k$$

The unit step sequence and shifted unit step sequence are shown in Fig. 3.27

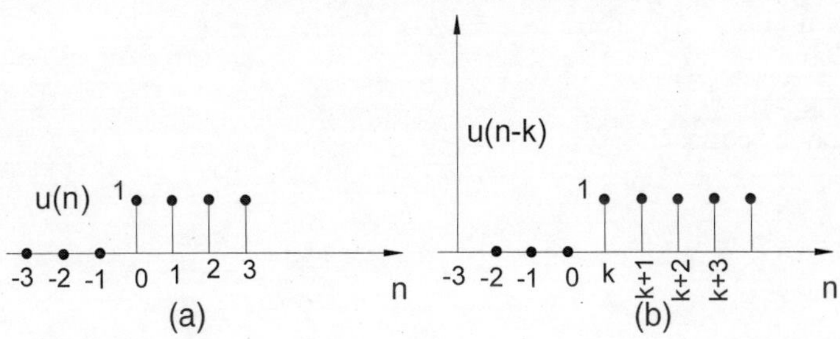

Fig. 3.27: Discrete time (a) Unit step function (b) Shifted unit step function

Unit ramp sequence

The unit ramp sequence is the discrete time sequence r(t) which starts with t = 0 and increases with time

$$r(n) = n, \text{ for } n \geq 0$$
$$= 0 \text{ for } n < 0$$

For shifted discrete time unit ramp sequence r(n-k) is given by :

$$r(n-k) = n, \text{ for } n \geq k$$
$$= 0 \text{ for } n < k$$

Fig 3.28 shows the unit ramp sequence

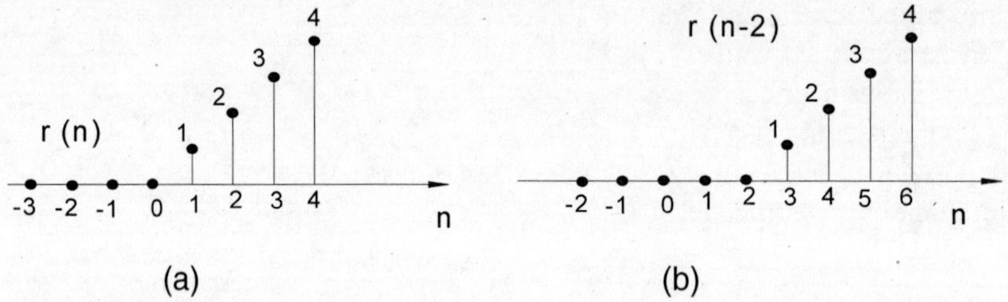

Fig. 3.28: Discrete time (a) Unit ramp sequence (b) Shifted ramp sequence

Unit parabolic sequence

The discrete time unit parabolic sequence p(n) is given by:

$$p(n) = \frac{n^2}{2}, \quad \text{for } n \geq 0$$
$$= 0 \text{ for } n < 0$$
$$\text{Or, } p(n) = \frac{n^2}{2}\, u(n)$$

For shifted discrete time unit parabolic sequence p(n -k) is given by :

$$p(n-k) = \frac{(n-k)^2}{2}, \text{ for } n \geq k$$
$$= 0 \text{ for } n < k$$
$$\text{Or, } \quad p(n\text{-}k) = \frac{(n-k)^2}{2}\, u(n)$$

Fig 3.29 shows the unit parabolic sequence

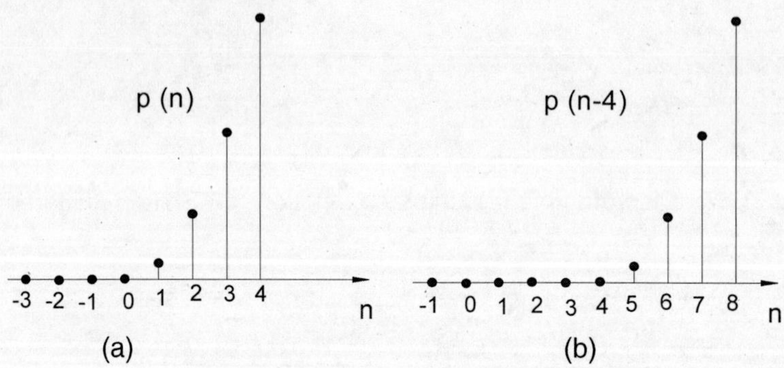

Fig. 3.29: Discrete time (a) Unit parabolic sequence (b) Shifted unit parabolic sequence

Unit impulse sequence

The discrete time unit impulse function δ(n) is defined as

$$\delta(n) = 1, \text{ for } n = 0$$
$$= 0 \text{ for } n \neq 0$$

For shifted discrete time unit impulse functionδ(n -k) is given by :

$$\delta(n - k) = 1, \text{ for } n = k$$
$$= 0 \text{ for } n \neq k$$

Fig 3.30 shows the unit impulse sequence

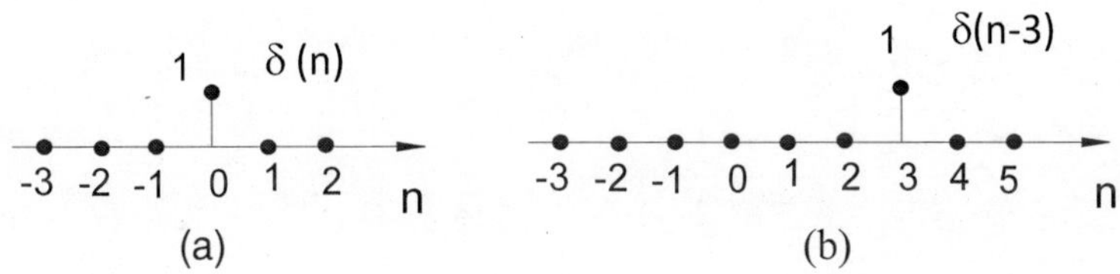

Fig. 3.30: Discrete time (a) Unit impulse sequence (b) Shifted unit impulse sequence

Sinusoidal sequence

The discrete time sinusoidal sequence is given by :

$$X(n) = A \sin(\omega n + \phi) \quad (3.8)$$

Where, A is the amplitude, ω is angular frequency, ϕ is phase angle in radians and n is an integer.

The period of the discrete time sinusoidal is $n = \frac{2\pi}{\omega} m$

Where n and m are integers

Fig 3.31 shows the unit impulse sequence

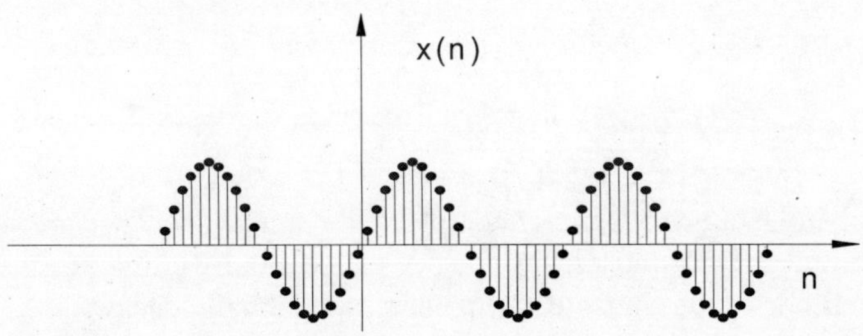

Fig. 3.31: Discrete time sinusoidal signal

Real Exponential Sequence

The discrete time real exponential sequence is defined as

$X(n) = a^n$ for all n

When a>1, the sequence grows exponentially as shown in Fig 3.32 (a)

When 0<a<1, the sequence decays exponentially as shown in Fig 3.32 (b)

When a < 0, the sequence takes alternating signs as shown in Fig 3.32 (c)

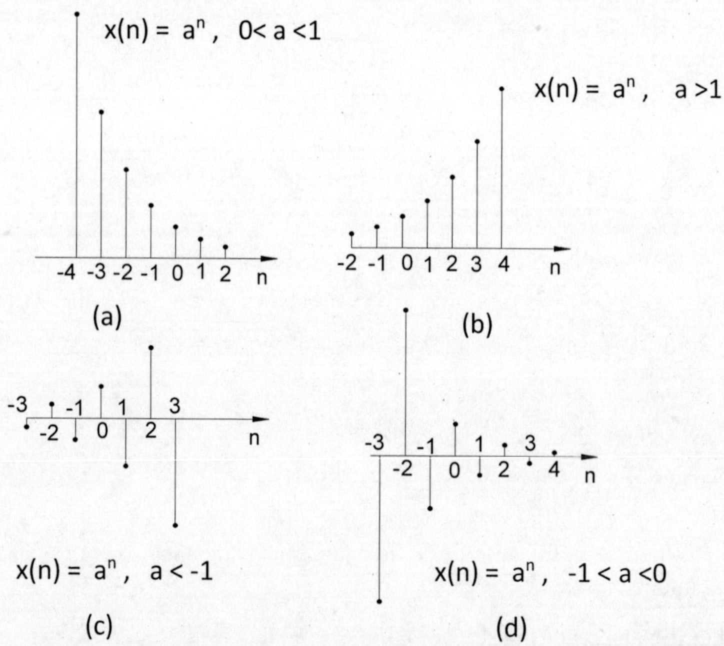

Fig. 3.32: Discrete time exponential signal

Complex exponential sequence

The discrete time complex exponential sequence is given by

$$X(n) = a^n e^{i(\omega_0 n + \phi)} \tag{3.9}$$

$$= a^n \cos(\omega_0 n + \phi) + i\, a^n \sin(\omega_0 n + \phi) \tag{3.10}$$

For $|a| = 1$, the real and imaginary parts of complex exponential sequence are sinusoidal.

For $|a| > 1$, the amplitude of the sinusoidal sequence exponentially grows

For $|a| < 1$, the amplitude of the sinusoidal sequence exponentially decays

Fig 3.33 shows the unit impulse sequence

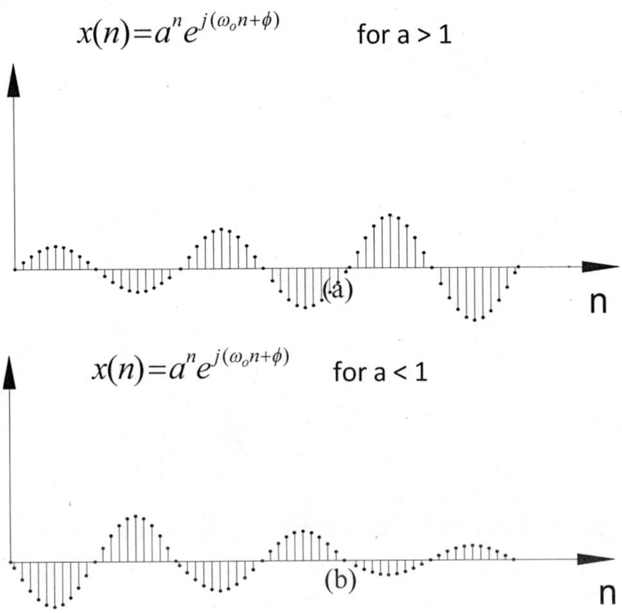

Fig. 3.33: Complex exponential sequence

Example 3.7

Ex. Find the following summations:

(a) $\sum_{-\alpha}^{\alpha} e^{2n}\,\delta(n-4)$

(b) $\sum_{-\alpha}^{\alpha} \delta(n-2)\cos 5n$.

Ans

(a) δ(n-4) is an impulse sequence,

δ(n - 4) = 1, for n = 4

= 0 for n ≠ 4

$$\sum_{-\alpha}^{\alpha} e^{2n}\,\delta(n-4) = e^{8} * 1 = e^{8}$$

(b) δ(n-2) is an impulse sequence,

δ(n - 2) = 1, for n = 2

= 0 for n ≠ 2

$$\sum_{-\alpha}^{\alpha} \delta(n-2)\cos 5n = 1.\cos(5.2) = \cos 10$$

3.4.3 Classification of Signal

Both continuous-time and discrete-time signals are classified into :

a) Deterministic and random signals
b) Periodic and non-periodic
c) Energy and power signals
d) Causal and non-causal signals
e) Even and odd signals

Deterministic and Random signals

A signal exhibiting some deterministic value of its amplitude and phase at any instant of time is a deterministic signal. Examples are:

a) Sinusoidal sequence, $x(n) = A \sin(\omega n + \phi)$
b) Exponential sequence, $x(n) = a^n$
c) Ramp sequence, $x(n) = n$, for $n \geq 0$
$= 0$ for $n < 0$

A deterministic signal can be completely represented by a mathematical equation.

A signal characterized by uncertainty about its occurrence is called a deterministic or random signal. A random signal can not be represented by a mathematical equation. The amplitude and phase of such a signal cannot be predicted in advance. A typical example is thermal noise.

Periodic and non-periodic

A signal which has a definite pattern and repeats itself after a regular interval of time is called a periodic signal and which does not repeat itself after any interval of time is called a non-periodic or aperiodic signal. A periodic signal has following general expression

$$x(n) = x(n+M) \text{ for all integers of n and M}$$

The smallest value of M for which the expression holds good is called the fundamental period of the signal.

If the above condition is not satisfied for any one value of n, the discrete time signal is aperiodic. Some examples of discrete time periodic/non-periodic signals are shown in Fig 3.34

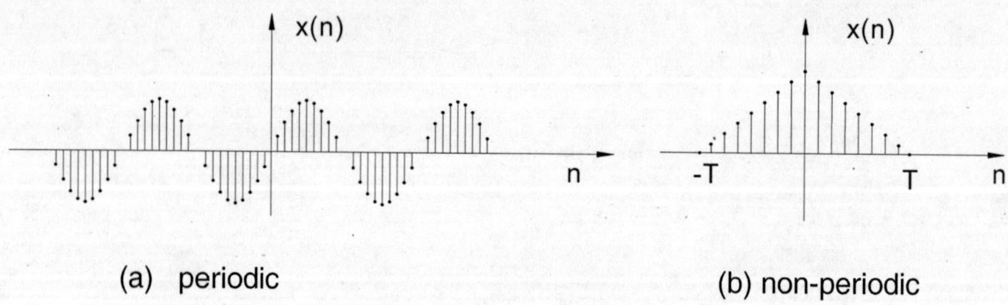

Fig. 3.34: Examples of periodic and non-periodic signals

Energy signal and Power signal

Signals can also be classified as energy signal and power signal.

The total energy E of a discrete time signal is given by:

$$E = \sum_{n=-\alpha}^{\alpha} |x(n)|^2$$

The average power P of a discrete time signal is given by :

$$P = \lim_{N \to \alpha} \frac{1}{2N+1} \sum_{n=-N}^{N} |x(n)|^2 \qquad (3.11)$$

Or, $P = \frac{1}{N} \sum_{n=0}^{n-1} |x(n)|^2$ for a digital signal with x(n) = 0 for n < 0

A signal is said to be an energy signal, if its energy over the interval (-α, α) is finite i.e. 0 < E <α. For a periodic signal the energy is either zero or infinite, hence the periodic signals are not energy signals.

A signal is said to be a power signal, if its power, P is finite, i.e. 0 <P<α. For a power signal, the total energy E is infinite (E = α). Periodic signals are the examples of power signals. Every bounded and periodic signal is a power signal.

Both energy and power signals are mutually exclusive, i.e. a discrete time signal can not simultaneously be an energy signal and a power signal.

Following signals are not energy signal or power signal:

x(n) = u(n); x(n) = n u(n); x(n) = n^2 u(n). For these signals neither P nor E are finite.

As a thumb rule it can be stated that for a signal whose amplitude is zero for $|n| \to \alpha$ is an energy signal. On the other hand, whose amplitude is not zero as $|n| \to \alpha$ is a power signal.

Causal and non-causal signal

A discrete time signal is causal if x(n) = 0 for n < 0, otherwise the signal is non causal. The time signal is said to be anti-causal if x(n) = 0 for n > 0. A causal signal does not exist for negative time and an anti-causal signal does not exist for positive time. A signal which exists in both positive and negative time is called a non-causal system. Thus u(n) is a causal signal but u(-n) an anti-causal signal, whereas x(n) = 2 for -3≤ n ≤ 3 is a non-causal signal.

Even signal and odd signal

A discrete time signal x(n) is said to be an even signal if it satisfies the following condition:

$x(n) = x(-n)$ for all values on *n*

Even signal is symmetrical about the vertical axis or time origin. A cosine sequence is an even signal.

A discrete time signal x(n) is said to be an odd signal if it satisfies the following condition:

$x(-n) = -x(n)$ for all values on *n*

Odd signal is anti-symmetrical about the vertical axis or time origin. A sine sequence is an odd signal.

Following are the examples of even and odd signal (Fig 3.35):

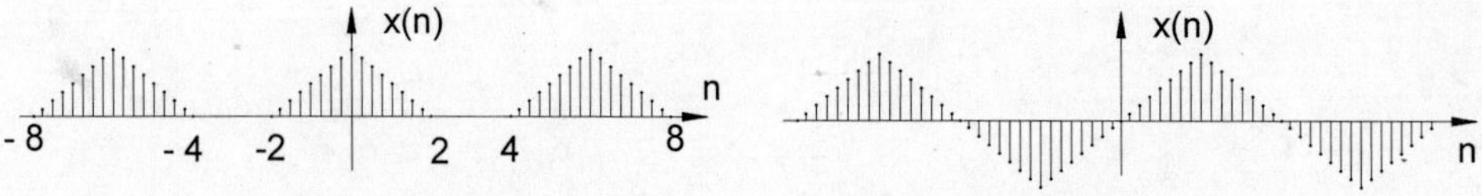

(a) Even sequence (b) Odd sequence

Fig. 3.35: Examples of even and odd sequences

Application of discrete time signal processing in Civil Engineering

The discrete time signal is used in different areas in Civil Engineering as discussed below:

a) Structural health monitoring : The discrete time signal is used in identifying damage, fatigue or anomalies in the structure.
b) Seismic analysis : the discrete time signal is used in the seismic analysis of ground and seismic effect on structures.
c) Traffic and transportation : the discrete time signal is used in the traffic and transportation engineering to assess traffic flow, signal processing, etc.
d) Environmental field : The discrete time signal is used in wind, rainfall, temperature, pollution field, etc.
e) Geotechnical engineering : The discrete time signal is used in geotechnical engineering like slope stability, settlement, liquefaction, etc.
f) Bridge monitoring : the discrete time signal is used in vibration analysis, potential weakness, crack propagation, etc.

3.5 NOISE

Noise is defined as unwanted and loud sound which is disruptive to hear. Figure 3.36 shows a signal carrying a noise.

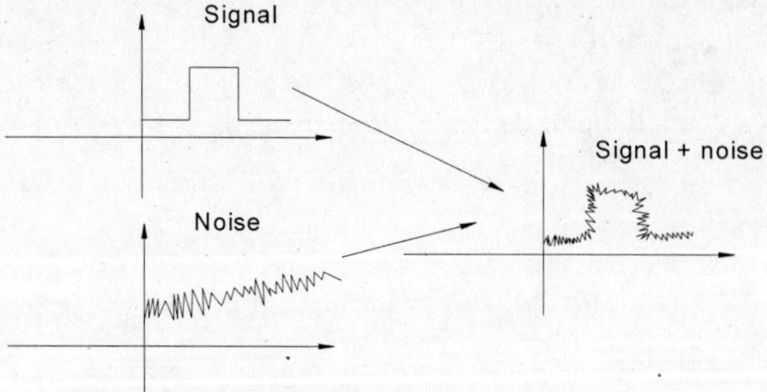

Fig. 3.36: Signal carrying a noise

Types of noise :

a) Continuous noise : This is a continuous flow of noise, for an example, noise produced by a machine uninterruptly.

b) Intermittent noise :Here noise increases and decreases intermittently, for example, aircraft movement, train passing across a place
c) Low frequency noise : low frequency noise from low noise diesel engine
d) Impulsive noise : the noise comes from sudden generation of noise like demolition of a building, etc.

The strength of signal with respect to the background noise is given by signal to noise ratio, S/N expressed in terms of decibel (db)

If the incoming signal strength is in microvolts V_s and the background noise level is in microvolts, V_n, the S/N is given by

$$S/N = 20\, log_{10} \frac{V_s}{V_n} \quad \text{in decibel}$$

If $V_s = V_n$, S/N = 0, it means that the signal is unreadable.

It is obvious that for any measurement system the S/N is to be very high. This can be achieved by increasing the signal level without any change in the noise level or by decreasing the noise level by some means.

Sources of noise :

The noise in the output may have following sources:

a) Generated noise
b) Conducted noise
c) Radiated noise.

Consider an input signal without any noise enters into an amplifier for a suitable output. The output signal will be amplifier gain times the input signal plus the noise. This noise is the generated noise. One of the possible reasons for generation of noise is that there are some internal components of the amplifier like resistors, capacitors, transistors, etc.

Noise may be generated during conduction of current in power wiring. The noise may be in the form of ripples, spikes or random deviations. This type of noise is the conducted noise. One of the most common sources of conducted noise is the 50 Hz power supply and the harmonics contained in it. The conducted noise can be reduces using suitable filters.

There may be electric or magnetic fields or disturbances in the environments around the amplifier because of which unwanted signals are radiated into the interior of the amplifier ad this is called radiated noise. Examples of radiated noise are electromagnetic impulses radiated from the ignition wiring of spark plugs. The radiated noise can be reduced by proper shielding.

Johnson Noise :The Johnson noise is the thermally generated noise. It increases with internal heating (I^2 R loss) or with an increase in ambient temperature. The noise power P_n is expressed as

$$P_n = k\, T\, \Delta f \qquad (3.12)$$

Where, k = Boltzman constant = 138 x 10^{-23} J/^{0}K

T = absolute temperature of resistor , ^{0}K

Δf = bandwidth of noise signal, Hz

The noise generation system can be represented by a voltage source E_n and an equivalent resistance R_n connected in series as shown in Fig. If the noise generator is connected with an external load resistance R_l, the noise energy will be transferred to the load. At maximum power transfer, $R_n = R_l$, the noise power is expressed as:

$$P_{nL} = \frac{E_n^2}{4R_n} \tag{3.13}$$

i.e. $$E_n = 2\sqrt{R_n P_{nL}} = 2\sqrt{R_n\ kT\ \Delta f} \tag{3.14}$$

Signal to noise ratio can also be expressed as the ratio of signal power to the noise power.

$$\frac{S}{N} = \frac{signal\ power}{noise\ power}$$

$$\frac{V_S^2/R}{V_n^2/R} = \frac{V_S^2}{V_n^2} \tag{3.15}$$

Where, V_s and V_n are the signal and noise voltage respectively.

Power Spectrum Density:

The power spectrum density S_n is defined as the noise power per unit of frequency bandwidth.

$$S_n = \frac{P_n}{\Delta f} = kT \tag{3.16}$$

The noise factor = F = $$\frac{S/N\ at\ input}{S/N\ at\ output} \tag{3.17}$$

Ex. An amplifier having a bandwidth 50 kHz has a noise power spectrum density input of 5 x 10^{-20} J. If the input resistance is 60 kΩ and the amplifier gain 100, what is the noise output voltage ?

Ans:

Given, power density function S_n = kT = 5 x 10^{-20} J

Input resistance R = 60 kΩ = 60 x 10^3Ω

Bandwidth Δf = 50 kHz = 50 x 10^3 Hz

Noise voltage input .$E_n = 2\sqrt{R_n\ kT\ \Delta f}$

$= 2\sqrt{5 * 10^{-20} * 60 * 10^3 * 50 * 10^3} = 24.5\ \mu V$

Noise voltage at the output =E_n x gain at amplifier

= 24.5 x 10^{-6} x 100 = 2.45 mV

Example 3.8

Ex. A low pass filter has an input S/N of 25. The input voltage is 5 mV. Calculate the noise voltage.

Ans:

$$\frac{S}{N} = \frac{signal\ power}{noise\ power}$$

$$=$$

$$25 = \frac{V_s^2}{V_n^2} = \frac{V_s^2}{V_n^2}$$

noise voltage: $V_n = \sqrt{\frac{V_s^2}{25}} = 1\ \text{mV}$

Example 3.9

3.5.1 Noise Removal

The noise can be removed by different techniques like:

a) Filtering : in this technique, the signal is passed through a filter which selectively attenuate certain selective frequencies and allows others to pass through. Examples are low pass filter, high pass filter, band pass, notch filters etc.
b) Adaptive filtering : this type of filter adjusts their parameters based on input signal characteristics. This is particularly suitable when the noise characteristics are not known in advance.
c) Wavelet transform: it decomposes signals into different frequency components. Thus noise with a given frequency can easily be eliminated. It is effective for both stationary and non-stationary noise
d) Statistical method: In this method, the noise in the signal is removed based on some statistical properties such as mean, variance, etc.
e) Denoising algorithm : Various denoising algorithms, such as the Wiener filter, Kalman filter and median filter use mathematical models to separate noise from the signal.
f) Frequency domain technique : in this method, Fourier transform or Fast Fourier Transform (FFT) are employed to analyse the signal in the frequency domain and then remove the noise components.
g) Machine Learning methods : machine learning algorithms such as neural network can be used to recognize the signal patterns and then remove them from the signal.

3.6 EXAMPLES ON STATISTICAL INFORMATION

Some examples on statistical information are given below.

3.6.1 Average value

The most probable value of measured variable is the arithmetic mean of the number of readings taken. The arithmetic mean is the average value of a large number of readings.

Definition : Sum of the observed values of a set divided by the number of observations is the average value.

If $X_1, X_2, X_3, X_4, \ldots, X_n$ are the N observed values, then the average value is given by:

$$\mu = \frac{X_1 + X_2 + X_3 + X_4 + \ldots + X_n}{N} = \frac{\sum X_i}{N} \tag{3.18}$$

Where, i = 1, 2, 3, ... , n

Example 3.10

Ex. A manufacturing company produces 2 different products. The production in units for each product in the last 10 days is as follows:

Product A: 15, 20, 25, 18, 22, 17, 21, 19, 24, 16

Product B: 12, 14, 13, 15, 17, 16, 18, 19, 20, 14.

Calculate the mean production for each product over the 10-day period

Ans.

Mean Production for Product A

= (15 + 20 + 25 + 18 + 22 + 17 + 21 + 19 + 24 + 16) / 10

= 197 / 10 = 19.7 units/day

Mean Production for Product B

= (12 + 14 + 13 + 15 + 17 + 16 + 18 + 19 + 20 + 14) / 10

= 158 / 10 = 15.8 units/day

When the data are arranged or given through frequency distribution, the formula for mean is given by:

$$\mu = \frac{f_1 X_1 + f_2 X_2 + f_3 X_3 + \ldots + f_n X_n}{f_1 + f_2 + f_3 + \ldots + f_n}$$

$$= \frac{\sum f_i X_i}{\sum f_i} \tag{3.19}$$

Where, i = 1, 2, 3, . . ., n

Ex. In an interview of eligible candidates, a total of 150 candidates appeared the interview. There were different age groups of the candidates. The distribution of age of different groups are shown below:

Age in years :	18	19	20	21	22	23	24	25
No. of candidates :	20	21	25	16	23	12	15	18

Calculate the average age of the candidates.

Ans.

Average age =

$$\frac{20*18+21*19+25*20+16*21+23*22+12*23+15*24+18*25}{20+21+25+16+23+12+15+18}$$

$$= \frac{3187}{150} = 21.25 \text{ years}$$

Example 3.11

If the data are given with p class intervals in the following form :

Class interval	Frequency
$X_1 - X_2$	f_1
$X_2 - X_3$	f_2
$X_3 - X_4$	f_3
. . .	. . .
$X_p - X_{p+1}$	f_p

Arithmetic mean $$\mu = \frac{f_1 Y_1 + f_2 Y_2 + f_3 Y_3 + ... + f_n Y_n}{f_1 + f_2 + f_3 + ... + f_n} \quad (3.20)$$

Where, Y_i is the midpoint of the class $X_i - X_{i+1}$

i.e. $$Y_i = \frac{X_i + X_{i+1}}{2}$$

Weighted mean :

If n number of data $X_1, X_2, X_3, ..., X_n$ have weights $w_1, w_2, w_3, ..., w_n$ respectively, then the weighted mean is given by

$$\mu = \frac{w_1 X_1 + w_2 X_2 + w_3 X_3 + ... + w_n X_n}{w_1 + w_2 + w_3 + ... + w_n} \quad (3.21)$$

Ex. In a city the sizes of the plots are divided into following classes with number of plots in every class. Find the average size of the plots

Size of plot in sq m	Mid point	Number of plots
0 - 150	75	100
150 - 300	225	80
300 - 450	375	60
450 - 600	525	40
600 - 750	675	20

Find the average size of the plots

Ans.

Average age =

$$\frac{100 * 75 + 80 * 225 + 40 * 525 + 20 * 675}{100 + 80 + 60 + 40 + 20}$$

$$= \frac{82500}{300} = 275 \text{ sq .m}$$

Example 3.12

3.6.2 Standard deviation

Standard variation is the square root of variance.

The variance is the average of the squares of the deviations taken from the mean.

If $X_1, X_2, X_3, \ldots, X_n$ are the data of N variables, as per the definition,

Variance =

$$\sigma^2 = \frac{\sum (X_i - \mu)^2}{N} \tag{3.22}$$

Where, i = 1, 2, 3, ... N

From the above equation we get

$$\sigma^2 = \frac{\sum X_i^2 - \left(\sum X_i\right)^2 / N}{N} \tag{3.23}$$

Where μ is the mean of the variables

If the data are given in the form of frequency distribution in which the variable X_i has a frequency distribution f_i,

The variance is given by :

$$\sigma^2 = \frac{\sum f_i (X_i - \mu)^2}{N} \tag{3.24}$$

For i = 1, 2, 3, ... N

and

$$N = \sum f_i \quad and \quad \mu = \frac{\sum f_i X_i}{N}$$

finally,

$$\sigma^2 = \frac{(\sum f_i X_i^2 - (\sum f_i X_i)^2 / N)}{N} \tag{3.25}$$

In the case of grouped data, the mid value is considered X_i and the above formula may be used.

However, the sample variance for a set $x_1, x_2, x_3, \ldots x_n$ of n observations is given by

$$\sigma^2 = \frac{\sum (x_i - \mu)^2}{n-1} \tag{3.26}$$

Or,

$$\sigma^2 = \frac{(\sum x_i^2 - (\sum x_i)^2 / n)}{n-1} \tag{3.27}$$

Where, i = 1, 2, 3, ..., n

and

$$\mu = \frac{\sum x_i}{n}$$

Ex. Find the variance and standard deviation of the following data on soil strength in N/mm^2

5, 12, 3, 18, 6, 8, 2, 10

Ans.

Mean $(x) = \frac{5+12+3+18+6+8+2+10}{8}$

Mean$(x) = \frac{64}{8} = 8$

Standard deviation $(\sigma) = \sqrt{\frac{\Sigma (x_i - x)^2}{n}}$

Variance $(\sigma^2) = \frac{\Sigma (x_i - x)^2}{n}$

Variance $(\sigma^2) =$

$\{(5-8)^2 + (12-8)^2 + (3-8)^2 + (18-8)^2 + (6-8)^2 + (8-8)^2 + (2-8)^2 + (10-8)^2\}/8$

Variance $(\sigma^2) = 194/8 = 24.25$

Standard deviation $(\sigma) = \sqrt{24.25} = 4.924$

Example 3.13

Ex. The following are the heights of nine persons in cm
165, 170, 172, 180, 159, 164, 160, 182, 196

Obtain the standard deviation of the height

Ans.

$$\text{Mean height } \mu = \frac{165+170+172+180+159+164+160+182+196}{9}$$

$$= 172$$

$(x_i - \mu)$ = -7 -2 0 8 -13 -8 -12 10 24

$(x_i - \mu)^2$ = 49 4 0 64 169 64 144 100 576

$\Sigma(x_i - \mu)^2$ = 49 + 4 + 0 + 64 + 169 + 64 + 144 + 100 + 576

= 1170

Variance = 1170 /8 = 146.25

Standard deviation = √ 146.25 = 12.09 cm

Example 3.14

3.6.3 Median

The median of a set of data is the middle number or centre value in the list. The median is also the number that is halfway into the set.

The data should be arranged in either ascending or descending order to find the median. The median is then obtained by picking the central value. If the total number of data is odd, then the median will be the central value. Still, if the total number of data is even, then there will be two central values, the median will be average of these two values.

Mathematically, if the total number of observations n is odd, the median is given by:

$$Median = (\frac{n+1}{2})th\ term \quad (3.28$$

If the total number of observations n is even, the median is given by:

$$Median = \frac{(\frac{n}{2})th\ term + (\frac{n+1}{2})th\ term}{2} \quad (3.29)$$

Ex. A group of BTech students obtained soil water holding capacity (in cm^3/g) at different locations in Bhubaneswar. The dataset is as follows:

15, 20, 23, 15, 17, 23, 20, 18, 15, 20, 25, 23, 15, 15, 17, 20, 25, 18, 20, 15

Find the mean soil water holding capacity.
Determine the median soil water holding capacity.

Ans.
Mean soil water holding capacity μ =

$$\frac{15+20+23+15+17+23+20+18+15+20+25+23+15+15+17+20+25+18+20+15}{20}$$

$$= \frac{396}{20} = 19.8 \quad cm^3/g$$

To find the median, first, arrange the data in ascending order:

15, 15, 15, 15, 15, 15, 17, 17, 18, 18, 20, 20, 20, 20, 20, 23, 23, 23, 25, 25

Since there are 20 data points (an even number), the median will be the average of the 10th and 11th values in the ordered list

Median = (20 + 20) / 2

Median = 20 cm^3/g

Example 3.15

3.6.4 Mode

The mode in statistics refers to a number in a set of numbers that appears the most. For an example, if a set of numbers contained the following digits :

1, 2, 5, 7, 9, 4, 5, 4, 5, 6, 8

The mode will be 5, as it occurs the most out of all the numbers in the set.

Ex. A group of B.Tech students obtained soil water holding capacity (in cm^3/g) at different locations in Bhubaneswar. The dataset is as follows:

15, 20, 23, 15, 17, 23, 20, 18, 15, 20, 25, 23, 15, 15, 17, 20, 25, 18, 20, 15

Find the mode of the data set.

Ans.

To find the mode (most frequent value), one can observe that the value '15' appears the most frequently in the dataset with a total of 7 times. Therefore, the mode is 15 cm^3/g.

Example 3.16

3.6.5 Range

The range of observations in the value obtained from the difference between the highest and the lowest values.

The formula for finding the range is given by :

Range = Highest observation – lowest observation.

To find the range, we need to arrange the given values or set of data in ascending order. From the above formula we then find the range.

Ex. Find the range of following observations:

15, 20, 23,17, 18, 25, 31, 48, 24, 52, 46, 35, 22, 30

Ans.

First arrange the observations in ascending order :

15, 17, 18, 20, 22, 23, 24, 25, 30, 31, 35, 46, 48, 52

The lowest value is 15

The highest value is 52

Range = 52 – 15 = 37

Example 3.17

Ex. Following are the marks of students in chemistry:

35, 38, 45, 65, 48, 68, 72, 92, 68, 75, 88, 39, 55, 68, 98, 83, 94, 89

Find the range of the marks

Ans.

First arrange the marks in ascending order :

35, 38,39, 45, 48, 55, 65, 68, 68, 68, 72, 75, 83, 88, 89, 93, 94, 98

The lowest value is 35

The highest value is 98

Range = 98 – 35 = 63

Example 3.18

UNIT SUMMARY

What have we learnt from module 3 ?

Module 3 is an introduction to the main course on Instrumentation for civil engineering applications. Following topics have been covered in this module :

1. Basic definitions of measurement, instrumentation and sensor.
2. Brief discussions on different types of sensors with figures.
3. Basic difference between a sensor and a transducer.
4. Different terminologies used in instrumentation.
5. Different types of instruments.
6. Interpretation of signal from a sensor.
7. Installation of different sensors in different civil structures namely metro station, elevated bridge and an underground tunnel.

EXERCISES

Ex. A garden contains 39 plants. The following plants were chosen at random, and their heights were recorded in cm: 38, 51, 46, 79, and 57. Calculate their heights' standard deviation

Ans.

Given that, Number of observations = 5
Hence, the mean of 5 observations is:
Mean = (38 + 51 + 46 + 79 + 57)/5 = 54.2
Now, the standard deviation is calculated as follows:
Standard Deviation, SD = $\sqrt{[(\Sigma(xi - \bar{x})^2) / (N\text{-}1)]}$
Now, substitute the values in the formula, we get

$$= \sqrt{\frac{(38-54.2)^2 + (51-54.2)^2 + (46-54.2)^2 + (79-54.2)^2 + (57-54.2)^2}{(5-1)}}$$

$$= \sqrt{\frac{962.8}{4}} = 15.51$$

Hence, the standard deviation (S.D) of their heights is 15.51 cm

Example 3.19

Ex. P(X) is the probability density, with X being a discrete random variable. Find the standard deviation of X.

X	1	2	3
P(X)	0.3	0.6	0.1

Ans.

Mean = $E(X) = \sum XP(X)$

Mean = $1(0.3) + 2(0.6) + 3(0.1) = 1.8$

And $E(X^2) = \sum X^2P(X) = 1^2(0.3) + 2^2(0.6) + 3^2(0.1)$

$E(X^2) = 3.6$.

Finding Variance:

Hence, Variance = $E(X^2) - [E(x)]^2$

Variance = $3.6 - (1.8)^2$

Variance = 0.36

Finding Standard Deviation:

Therefore, the standard deviation = $\sqrt{(0.36)} = 0.6$

Example 3.20

Ex. P(X) is the probability density, with X being a discrete random variable. Find the standard deviation of X.

X	1	2	3
P(X)	0.3	0.6	0.1

Ans.

Mean = $E(X) = \sum XP(X)$

Mean = $1(0.3) + 2(0.6) + 3(0.1) = 1.8$

And $E(X^2) = \sum X^2P(X) = 1^2(0.3) + 2^2(0.6) + 3^2(0.1)$

$E(X^2) = 3.6$.

Finding Variance:

Hence, Variance = $E(X^2) - [E(x)]^2$

Variance = $3.6 - (1.8)^2$

Variance = 0.36

Finding Standard Deviation:

Therefore, the standard deviation = $\sqrt{(0.36)} = 0.6$

Example 3.21

Ex. Using the actual mean method, calculate the standard deviation for the data 3, 2, 5, & 6

Ans.
Given data observations are 3, 2, 5, and 6.
The number of observations = 4.
Finding Mean:
Now, the mean of the given data is:
Mean = (3+2+5+6)/4
Mean = 16/4
Mean = 4.
Finding Variance:
The squared differences from mean = $(4-3)^2 + (2-4)^2 + (5-4)^2 + (6-4)^2 = 10$
Variance = Squared differences from mean/Total number of data
Variance =10/4
Therefore, variance =2.5
Finding Standard Deviation:
S.D = $\sqrt{2.5}$
S.D = 1.58.
Therefore, the standard deviation is 1.58

Example 3.22

Ex. Calculate the range of the first 10 prime numbers.

Ans.
The first 10 prime numbers are 2, 3, 5, 7, 11, 13, 17, 19, 23, 29.
Here, the lowest data value is 2 and the highest data value is 29.
We know that the formula to calculate the range is the highest data value minus the lowest data value.
Thus, Range = Highest data value – Lowest data value
Range = 29 – 2
Range = 27.
Therefore, the range of the first 10 prime numbers is 27

Example 3.23

Ex. Determine the highest value in the data set, if the range equals 40 and the lowest value should be equal to 6.

Ans.
Given that the lowest value in the dataset is 6.
And, range = 40.
As we know, Range = Highest data value – Lowest data value.
Now, substitute the values in the formula, and we get
40 = Highest value – 6
Therefore, the highest value = 40 + 6
Highest value = 46.
Hence, the highest value in the dataset is 46.

Example 3.24

Ex. Determine the range of the bearing capacity (N/mm^2) of the soil of a locality: 12, 21, 9, 45, 32, and 40.

Ans.
Given data: 12, 21, 9, 45, 32, 40 (N/mm^2)
Now, the ascending order of given data values is
9, 12, 21, 32, 40, and 45 (N/mm^2)
Hence, the lowest value = 9 and highest value = 45.
Range = Maximum value – Minimum value
Range = 45 – 9
Range = 36
Hence, the range of the given data values = 36 (N/mm^2)

Example 3.25

MULTIPLE CHOICE QUESTIONS

Q.1 :LVDT is a

a) capacitive transducer b) inductive transducer

c) resistive transducer d) none of them

Q.2 : LVDT windings are wound on

a) steel sheets b) aluminium

c) irond) copper

Q.3 : Piezoelectric crystals are used for measuring changes in

a) static condition b) none of above

c) both static and dynamic d) dynamic condition

Q.4 : Piezoelectric crystals produce emf

a) when external mechanical force is applied b) when heat is applied

c) when radio waves stimulate the crystal c) when the external magnetic field is applied

Q 5 : Piezoelectric crystals are used for the measurement of

a) sound b) velocity

c) Temperature d) none of the above

Q.6 : Capacitive transducers are normally used for the following measurement

a) static b) dynamic

c) both static and dynamic d) transient

Q.7 : The gauge factor of a strain gauge is given by:

a) $GF = \frac{\Delta l/l}{\Delta R/R}$ b) $GF = \frac{\Delta l/l}{\Delta d/d}$

c) $GF = \frac{\Delta l/R}{\Delta R/l}$ d) $GF = \frac{\Delta R/R}{\Delta l/l}$

Q.8 : The gauge factor and the Poisson's ratio are related with

a) $GF = 2 + \mu$ b) $GF = 2\mu$

c) $GF = 1 + 2\mu$ d) $GF = (1 + \mu)/2$

Q.9 : LVDT can be used for

a) vibration measurement b) force measurement in a beam

c) angular velocity measurement d) load measurement in a column

Q.10 : An LVDT has an output in the form of

a) linear displacement of the core b) pulse

c) rotary movement of core d) none of the above

Q.11 A strain gauge should have a low

a) resistant temperature coefficient b) resistance

c) gauge factor d) all of the above

Q.12 Seismic transducer is used for the measurement of

a) Linear velocity b) acceleration

c) angular velocity d) pressure

Q.13 A load cell is essentially a

a) inductive transducer b) thermister

c) resistive potentiometer d) strain gauge

Q.14 A load cell is an electro-mechanical device and is widely used for measurement of

a) static forces b) dynamic forces

c) temperature d) both (a) and (b)

Q.15 Which of the following can be used for pressure measurement

a) thermometer b) pyrometer

c) piezoelectric crystal d) bolometer

Q.16 A Wheatstone bridge circuit using strain gauges can be used for measuring

a) static strains b) dynamic strains

c) both (a) and (b) d) none of the above

Answers of Multiple Choice Questions
1(b), 2(c), 3(d), 4(a), 5(c), 6(c), 7(d), 8(c), 9(b), 10(a), 11(a), 12(b), 13(d), 14(d), 15(c), 16(c)

SHORT AND LONG ANSWER TYPE QUESTIONS

Q1. Explain briefly the terms: accuracy, errors and correction.

Q2. Discuss the working principle of LVDT

Q3. How does strain gauge load cell work ?

Q4. How does the torque sensor work ?

Q5. Discuss briefly the working of the accelerometer

Q6. What are the different types of piezometers ?

Q7. Explain the working principle of the pneumatic piezometer.

Q8. How can earth pressure be measured ?

Q9. Discuss different aspects of the installation of sensors.

Q10. What are the different types of sensors ?

Q11. Discuss the working principle of piezoelectric sensor

Q12. How does a capacitive sensor work ?

Q13. How are uncertainties estimated ?

Q14. How does an extension meter work ?

Q15. What is a load cell ?

Q16. Discuss how to measure the gauge factor for a half bridge Wheatstone bridge circuit ?

Q17. Obtain the change in output voltage for a full bridge Wheatstone bridge circuit.

Q18. What are the different techniques of interpretation ?

Q19. Discuss the working principle of the inclinometer.

Q20. What are the different types of strain gauge rosettes ?

Q21. What are the different types of discrete signals ?

Q22. Mention different classes of signals.

Q23. How is noise removed from a signal ?

REFERENCES AND SUGGESTED READINGS

1. Alan S Morris (2001). Measurement and Instrumentation Principles, 3rd/ed, Butterworth Hienemann.
2. Beeby, S., Ensell, G., Kraft, M., and White, N (2004). MEMS Mechanical Sensors. Artech House, Inc., Boston, London
3. David A. Bell (2007). Electronic Instrumentation and Measurements 2nd /ed, Oxford Press.
4. Dunnicliff, J. (1982). Geotechnical Instrumentation for monitoring field performance, Report 89. Transportation Research Board, National Research Council, Washington, D.C
5. Fraden, J. (2003). Handbook of modern sensors. Springer-Verlag, New York.
6. https://www.automatizari-cladiri.ro/fise-tehnice/A6V10366334.pdf
7. Ilya Gertsbakh (2010). Measurement Theory for Engineers, Springer
8. Internet sources

9. Judy, J.W., R.S. Muller,and H.H. Zappe, "Magnetic Microactuation of Polysilicon Flexure Structures," Tech. Dig. Solid State Sensor and Actuator Workshop, Hilton Head, SC, 1994, pp 43-48
10. Kantardzic M, " Data Mining: Concepts, Models, Methods, and Algorithms, New York, Wiley, 2003.
11. Rajput, R. K. (2017). "Mechanical Measurements and Instrumentation" . S.K.Kataria& Sons, New Delhi.
12. S. Tumanski (2006). Principle of Electrical Measurement, Taylor & Francis
13. Tanigawa H., Ishihara T., Hirata M., Suzuki K. (1985). MOS integrated silicon pressure sensor IEEE Trans Electron Devices ED -32 (7) : 1191-1195

KNOW MORE

For more information related to this topic scan the QR code.

OR

Type this link in your browser

https://digimat.in/nptel/courses/video/113101096/L01.html

4 Frequency Domain Signal Processing and Analysis

UNIT SPECIFICS

Through this unit we have discussed the following aspects:

- *Basic definition of frequency domain signal processing;*
- *The importance of frequency domain analysis and its principles;*
- *Applications of frequency domain analysis;*
- *The physical processes based on analysis of sensor data;*
- *Basic concepts in frequency domain signal processing and analysis a) Fourier Transform b) FFT (Fast Fourier Transform);*
- *How do frequency domain signal processing techniques transform the signal from the time domain to the frequency domain using Fourier transformation?*
- *Implementation of FFT in MATLAB and Mathcad;*
- *Concepts of Noise reduction with filters, Leakage, and Frequency resolution with application*

The practical applications of the topics are discussed solving to generate further curiosity and creativity as well as enhance problem solving capacity.

Besides giving many multiple-choice questions as well as questions of short and long answer types marked in two categories following lower and higher order of Bloom's taxonomy, assignments through several numerical problems, a list of references and suggested readings are given in the unit so that one can go through them for practice.

RATIONALE

This unit on frequency domain signal processing helps students get a basic idea about the importance of frequency domain analysis, and its application in Civil Engineering. It explains how frequency domain signal processing techniques transform the signal from the time domain to the frequency domain utilising Fourier transformation, application of FFT in MATLAB and Mathcad. Basic ideas about the different noise reduction filters, reason for leakage, and frequency resolution. Some example problems are also added along with some MCQs at the end of the chapter for better understanding of the topic mentioned above from an application point of view.

PRE-REQUISITES

Mathematics: General Mathematics (Class XII)

Physics: Engineering Physics (First year, B.Tech.)

UNIT OUTCOMES

The list of outcomes of this unit is as follows:

U4-O1: Describe basic definitions of frequency domain signal processing

U4-O2: Applications of frequency domain analysis

U4-O3: Basic concepts in frequency domain signal processing and analysis

U4-O4: Implementation of FFT in MATLAB and Mathcad

U4-O5: Concepts of Noise reduction with filters, Leakage, and Frequency resolution with application

Unit-4 Outcomes	**EXPECTED MAPPING WITH COURSE OUTCOMES** *(1- Weak Correlation; 2- Medium correlation; 3- Strong Correlation)*							
	CO-1	**CO-2**	**CO-3**	**CO-4**	**CO-5**	**CO-6**	**CO-7**	**CO-8**
U4-O1	-	-	-	-	1	-	1	-
U4-O2	1	1	1	-	-	2	1	1
U4-O3	-	-	-	-	1	1	-	-
U4-O4	1	-	-	-	-	1	-	-
U4-O5	3	-	3	-	-	-	1	1

4.1 FREQUENCY DOMAIN ANALYSIS OF SIGNALS

Frequency domain signal processing is used to analyse and manipulate signals based on their frequency characteristics. It involves transforming signals from the time domain to the frequency domain, where they are represented in terms of frequency components. This transformation allows for a deeper understanding of the signal's frequency content. In the frequency domain, various operations such as spectrum analysis, filtering, modulation, and equalization can be performed. These operations provide insights into the distribution of frequencies, enable the removal or modification of specific frequency components, and help compensate for frequency-dependent distortions. Frequency domain signal processing has applications in telecommunications, audio, image, and control systems.

The Fourier transform is widely used in frequency domain analysis as it converts signals from the time domain to the frequency domain. This transformation allows for identifying the distinct frequency components within the signal. By decomposing the signal into a combination of sinusoidal components with different frequencies, amplitudes, and phases, the Fourier transform provides valuable insights into the signal's frequency content. Once the signal is in the frequency domain, various operations can be applied for analysis and processing. These operations may include spectrum analysis, filtering, convolution, equalization, and spectral analysis. Each operation leverages the frequency domain representation to extract meaningful information and manipulate the signal based on its frequency characteristics. Some of the common tasks include:

- **Spectrum analysis**: It is a valuable technique that allows for examining a signal's frequency content. By analysing the spectrum, one can gain insights into the strength and distribution of different frequencies within the signal. This analysis is crucial for understanding the dominant frequency components and detecting any unwanted noise or interference that may be present. The ability to identify specific frequencies causing disruptions or affecting signal quality is particularly important. By pinpointing these problematic frequencies through spectrum analysis, researchers and engineers can implement targeted mitigation techniques. These techniques can help minimize the impact of unwanted frequencies, reduce noise or interference, and ultimately enhance the overall quality and reliability of the signal. Spectrum analysis plays a significant role in various applications where precise knowledge of frequency components is essential.

- **Filtering**: Frequency domain filtering is a powerful technique used to eliminate or weaken specific frequency components within a signal selectively. By applying filters in the frequency domain, unwanted noise or interference can be effectively removed. Different filters, such as low-pass, high-pass, bandpass, or notch filters, can be employed to achieve specific filtering objectives. A low-pass filter allows frequencies below a specific cutoff value to pass through while attenuating higher frequencies. This filter is useful for removing high-frequency noise. Conversely, a high-pass filter permits frequencies above a specific cutoff value while attenuating lower frequencies. It is effective in eliminating low-frequency noise. Bandpass filters allow a specific range of frequencies to pass through while attenuating others. They are utilized when isolating a specific frequency band is desired. Notch filters, on the other hand, suppress a narrow range of frequencies and are commonly used for removing specific interference signals. By employing frequency domain filtering techniques, unwanted frequency components can be selectively reduced or eliminated, leading to improved signal quality and enhanced detection or analysis of desired signals.

- **Convolution**: Convolution in the frequency domain is mathematically equivalent to multiplication in the time domain. By multiplying the frequency representations of two signals, the result is equivalent to the convolution of their time domain counterparts. This property is known as the convolution theorem. This operation has various applications in Signal processing. One common application is signal smoothing, where the convolution of a signal with a smoothing kernel in the frequency domain can effectively reduce noise or irregularities in the time domain. Deconvolution, which involves reversing the effects of convolution, can also be performed by multiplication in the frequency domain. Modulation, a technique used to modify a signal by shifting its frequency components, can also be achieved through frequency domain convolution. By multiplying the frequency representation of the signal with a modulation function, the resulting frequency domain convolution generates the desired frequency-shifting effect in the time domain. By leveraging the properties of convolution in the frequency domain, signal processing tasks such as smoothing, deconvolution, and modulation can be efficiently performed, enabling various applications in audio and image processing, communications, and many other fields.

- **Equalization**: Frequency domain equalization is a technique used to mitigate the frequency-dependent distortions introduced by transmission channels or other systems. Its purpose is to restore the original shape of a signal by modifying its frequency response. This is typically achieved through inverse filtering in the frequency domain. When a signal traverses a transmission channel or encounters a system, it may experience alterations in its frequency characteristics. These alterations can cause certain frequencies to be attenuated or emphasized, resulting in distortions in the signal's shape. Frequency domain equalization aims to counteract these distortions by applying modifications to the frequency response of the signal. Inverse filtering is a common approach used in frequency domain equalization. It involves estimating the frequency response of the channel or system and applying the inverse of this response to the distorted signal. This counteracts the frequency-dependent distortions, effectively restoring the original shape of the signal by compensating for the frequency-dependent distortions, frequency domain equalization helps to improve the fidelity and quality of transmitted signals. It finds applications in various areas, such as telecommunications, audio processing, and wireless communication systems, where correcting frequency-dependent impairments is crucial for achieving accurate and reliable signal transmission.

- **Spectral analysis**: Spectral analysis techniques, including power spectral density estimation, offer valuable insights into power distribution across different frequency bands within a signal. By analyzing the power spectrum, information about the energy distribution and strength of various frequency components can be obtained. This analysis is useful in applications such as audio and speech processing, where understanding the frequency content and energy distribution is essential for tasks like noise reduction, speech enhancement, or audio compression. In vibration analysis, spectral analysis helps identify resonant frequencies and modes of vibration, aiding in structural health monitoring and fault detection. Additionally, in system identification, spectral analysis assists in determining the frequency response characteristics of a system, enabling the estimation of system parameters. By examining the power spectral density, researchers and engineers gain valuable information about signals' dominant frequencies, bandwidths, and energy contributions. This knowledge facilitates informed decision-making and enables the development of effective signal processing algorithms for a wide range of applications.

Frequency domain signal processing and analysis play a critical role in various fields where understanding and manipulating signal frequency content are crucial. Some of these fields include:

- **Audio and Speech Processing**: Frequency domain techniques are widely utilized for noise reduction, speech enhancement, audio compression, and audio effects processing. Desired modifications and improvements can be achieved by analysing and manipulating the frequency components of audio signals.
- **Telecommunications**: In telecommunications systems, frequency domain analysis is fundamental for signal modulation, channel equalization, error correction, and data compression. Understanding the frequency characteristics of signals is essential for reliable and efficient communication.
- **Image Processing**: Frequency domain analysis is utilized in image processing for tasks such as image enhancement, filtering, compression, and feature extraction. Transformations like the Fourier transform and its variants allow for the examination and manipulation of image frequency components.
- **Control Systems**: Frequency domain techniques are extensively employed in control systems engineering. They enable the analysis and design of control systems by characterizing the frequency response of system components, stability analysis, and controller design using techniques like the Bode plot, Nyquist plot, and frequency response methods.
- **Signal Analysis and Scientific Research**: Frequency domain analysis provides valuable insights into signal characteristics, making it applicable in various scientific research fields. It aids in analysing of signals in areas such as biomedical research, environmental monitoring, geophysics, and more.

By utilizing frequency domain signal processing and analysis, researchers and engineers can better understand signal behaviour, manipulate signals for desired outcomes, and develop advanced techniques and algorithms for signal processing tasks in diverse domains.

A voice waveform in the time and frequency domain is shown in Fig 4.1

(a) Time domain

(b) Frequency domain

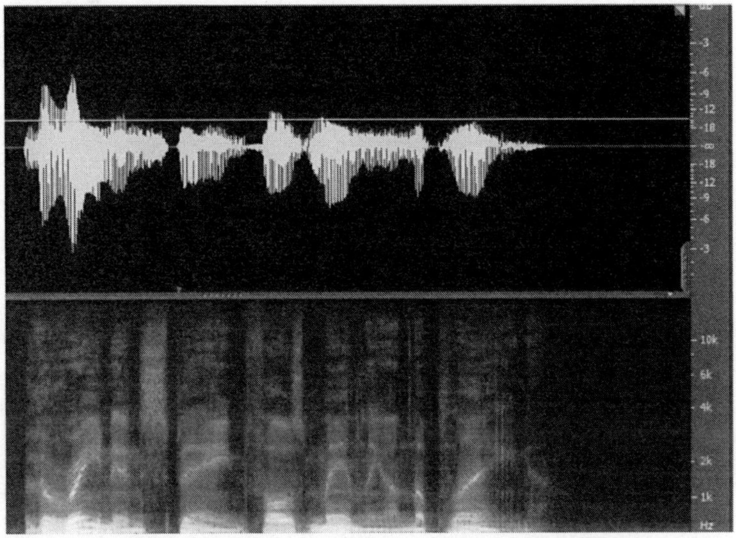

Fig 4.1 A voice waveform in the Time domain and frequency domain (After Santamarina et al., 2005)

4.1.1 Applications of frequency domain signal processing and analysis

Frequency domain signal processing and analysis has numerous applications across various fields which include:

- **Audio and Speech Processing**: Frequency domain analysis plays a fundamental role in applications within the realm of audio and speech processing. It enables us to delve into the frequency content of audio signals, facilitating various tasks encompassing speech recognition, music analysis, and audio compression. In speech recognition, frequency domain analysis proves invaluable in identifying phonetic elements and distinguishing between different speech sounds. Techniques like the Fourier transform and cepstral analysis extract pertinent features and model speech patterns. By examining the frequency components of speech signals, we can uncover significant information crucial for accurate recognition and understanding of spoken words. Moving to music analysis, frequency domain techniques offer a means to undertake essential tasks such as pitch detection, melody extraction, and timbre analysis. By decomposing music signals into their constituent frequency components, we gain insights into the music underlying melodic and harmonic elements. This aids in pitch estimation, allowing us to determine the fundamental frequency and extract the melodic structure. Moreover, timbre analysis provides a deeper understanding of the sonic characteristics and instrumentation in a piece of music. Furthermore, frequency domain analysis plays a pivotal role in audio compression Algorithms, exemplified by widely used formats such as MP3. These algorithms capitalize on frequency domain analysis to identify and eliminate perceptually insignificant components within the audio signal. By exploiting the psychoacoustic properties of human hearing, frequency components less likely to be detected by the human ear can be efficiently removed. This reduction in data allows for the compression of audio files while still maintaining an acceptable level of audio quality. The Fourier transform serves as a cornerstone of frequency domain analysis, enabling the conversion of signals from the time domain to the frequency domain. This transformation allows for the representation of signals in terms of their constituent frequencies and provides a valuable perspective on their spectral characteristics. By expressing signals in the frequency domain, we gain access to a wealth of information regarding the distribution of frequencies and their respective amplitudes. Cepstral analysis is another technique utilized in frequency domain analysis, that is particularly relevant to speech processing. It involves a two-step transformation that combines the frequency spectrum of a signal with its corresponding logarithm. This process allows for the separating the source and filter characteristics of the speech signal, enabling a deeper understanding of the underlying vocal tract properties and facilitating tasks like speaker recognition and speech synthesis. The significance of frequency domain analysis is not limited to audio and speech processing alone. It extends to other fields, including telecommunications, image processing, control systems, and scientific research. In telecommunications, frequency domain analysis plays a critical role in tasks such as modulation, channel equalization, and error correction. It aids in the reliable transmission and reception of signals by understanding their frequency characteristics and mitigating distortions that may occur during transmission. Within image processing, frequency domain analysis enables tasks such as image enhancement, filtering, compression, and feature extraction. By transforming images into the frequency domain, we can manipulate their frequency components to achieve desired effects. Techniques like the Fourier transform and its variants are employed to analyse the frequency content of images, leading to applications in fields like medical imaging, computer vision, and digital photography. Control systems engineering heavily relies on frequency domain analysis to characterize the frequency response of system components, assess stability, and design controllers. Techniques like Bode plots, Nyquist plots, and frequency response methods allow for the analysis and design of control systems in diverse applications ranging from industrial automation to aerospace systems. Additionally, frequency domain analysis finds application in scientific research, aiding in the analysis of signals in fields such as biomedical research, environmental monitoring, geophysics, and beyond. By examining the frequency content of signals, researchers can uncover valuable insights and patterns, leading to advancements in scientific understanding and discoveries. In summary, frequency domain analysis is a fundamental tool in audio and speech processing, enabling speech recognition, music analysis, and audio compression. By investigating the frequency content of signals, we can extract meaningful features, model patterns, and achieve efficient data compression. Moreover, frequency domain analysis extends its impact to diverse fields, including telecommunications, image processing, control systems, and scientific research, where understanding and manipulating signal frequency characteristics are paramount. Its versatility and wide-ranging applications highlight its significance in modern signal processing and analysis.

- **Telecommunications**: Frequency domain signal processing plays a vital role in various aspects of telecommunications, particularly in signal modulation, channel equalization, and error correction. Modulation techniques, including amplitude modulation (AM) and frequency modulation (FM), involve manipulating the frequency components of signals to transmit information efficiently. By analysing modulated signals in the frequency domain, the demodulation process becomes more effective, allowing for the recovery of the original data. In wireless communication systems, frequency domain analysis is utilized for tasks like channel estimation and equalization. By characterizing the frequency response of wireless channels, equalization algorithms can compensate for distortions introduced during signal transmission. This improves the reliability and quality of the received data, especially in environments with fading or multipath propagation effects. Frequency domain equalization techniques, such as adaptive equalizers, ensure that the transmitted signals are correctly received by mitigating the adverse impact of channel impairments. Error correction coding techniques, such as forward error correction (FEC), heavily rely on frequency domain analysis. FEC enables the detection and correction of transmission errors, ensuring data integrity. By leveraging frequency domain analysis, error correction codes can be designed to detect and correct errors during signal transmission. Techniques like Reed-Solomon coding and turbo coding exploit the frequency domain properties to efficiently encode and decode data, enhancing system data transmission's robustness and improving the communication system overall reliability. Overall, frequency domain signal processing is an integral part of telecommunications systems, playing a critical role in modulation, channel equalization, and error correction. Telecommunications systems can achieve efficient and reliable data transmission through the analysis and manipulation of signal frequency components, ensuring high-quality communication in various environments and applications.

- **Image Processing**: In image processing, frequency domain techniques play a significant role in various tasks such as image enhancement, compression, and restoration. These techniques leverage the frequency content of images to achieve desired outcomes. Frequency domain filtering is used for image enhancement to eliminate noise, blurring, or other undesired artifacts. Filters such as low-pass, high-pass, and bandpass filters can be applied to modify the frequency components of an image, resulting in improved image quality. Image details can be improved by selectively altering the frequency content, while reducing unwanted disturbances. Image compression algorithms, like JPEG, take frequency domain analysis to exploit redundancies in image data. Transform coding techniques, such as the Discrete Cosine Transform (DCT), convert images into their frequency representation. This representation allows for efficient compression by quantizing and encoding the frequency coefficients, thereby reducing the storage size while maintaining acceptable image quality. Frequency domain analysis is also utilized in image restoration to address image degradations caused by factors such as motion blur or noise. By analysing the frequency content of degraded images, restoration algorithms can estimate and compensate for the distortions, leading to improved image quality. In control systems engineering, frequency domain analysis plays a crucial role in stability analysis, controller design, and system identification. Frequency response analysis is employed to understand the system's behaviour as a function of frequency. It aids in assessing stability margins, such as gain margins and phase margins, ensuring reliable and robust system operation. Bode and Nyquist plots are commonly used graphical representations for frequency domain analysis in control systems. Controller design techniques, like proportional-integral-derivative (PID) control, often incorporate frequency domain analysis. By examining the system's frequency response and characteristics, optimal controller parameters can be determined to achieve desired performance specifications. System identification techniques also rely on frequency domain analysis to estimate the parameters of a system from input-output data. Frequency response functions and transfer function models are employed to represent the system dynamics accurately, facilitating precise modelling and control. In summary, frequency domain techniques are used extensively in image processing for image enhancement, compression, and restoration tasks. These techniques leverage the frequency content of images to achieve desired effects. In control systems engineering, frequency domain analysis is utilized for stability analysis, controller design, and system identification, providing valuable insights into system behaviour, and enabling optimal performance.

- **Biomedical Signal Processing**: Frequency domain analysis plays a critical role in biomedical signal processing, encompassing various applications such as electroencephalography (EEG), electrocardiography (ECG), and medical imaging. In EEG analysis, frequency domain techniques are essential for identifying distinct brain wave patterns associated with different mental states or pathologies. Power spectral density estimation provides valuable information about the power distribution across different frequency bands in the EEG signal, aiding in the understanding and of characterizing brain activity. In ECG analysis, frequency domain analysis proves beneficial in detecting abnormalities and arrhythmias. Techniques like Fourier analysis and wavelet transforms allow for identifying specific frequency components associated with heart rate variations or cardiac anomalies. By examining the frequency content of the ECG signal, clinicians and researchers can gain insights into the underlying cardiovascular dynamics and diagnose potential cardiac conditions. Frequency domain methods also have extensive applications in medical imaging. Fourier-based MRI reconstruction techniques enable rapid image reconstruction from acquired data, facilitating real-time imaging, and reducing scanning times. Frequency domain analysis, such as Fourier imaging and wavelet analysis, is further employed for tasks like image denoising, feature extraction, and image registration in various medical imaging modalities. These techniques leverage the frequency information embedded within the image data to enhance image quality, extract relevant features, and align images for accurate analysis and diagnosis. The applications of frequency domain signal processing and analysis extend beyond biomedical signal processing. Its versatility and effectiveness make it a valuable tool in other fields also. In audio and telecommunications, frequency domain techniques enable efficient modulation, equalization, and error correction. In image processing, frequency domain methods contribute to image enhancement, compression, and restoration. Control systems benefit from frequency domain analysis for stability assessment, controller design, and system identification. In conclusion, frequency domain analysis role is crucial in biomedical signal processing, aiding in EEG analysis, ECG analysis, and medical imaging. The applications encompass brain wave pattern identification, cardiac anomaly detection, fast MRI reconstruction, image denoising, feature extraction, and image registration. Additionally, frequency domain analysis finds application in audio, telecommunications, image processing, control systems, and other fields. Its versatility and effectiveness make it a valuable tool for understanding and manipulating signals in various domains.

4.1.2 How frequency domain signal processing techniques transform signal from time domain to frequency domain using Fourier Transformation

The Fourier transform is the primary tool for transforming a signal from the time domain to the frequency domain in frequency domain signal processing. The Fourier transform decomposes a signal into its constituent frequency components, revealing the amplitudes and phases associated with each frequency. Here's a step-by-step explanation of how the Fourier transform is applied:

- Define the Signal: Start with a signal typically represented as a function of time, denoted as x(t). This signal can be a continuous-time signal or a discrete-time signal, depending on the nature of the data.
- Apply the Fourier Transform: The Fourier transform converts the signal from the time domain to the frequency domain. Different variants of the Fourier transform depend on whether the signal is continuous or discrete. The continuous-time Fourier transform (CTFT) is used for continuous-time signals, while the discrete Fourier transform (DFT) or its fast algorithm, the fast Fourier transform (FFT), is used for discrete-time signals.

For continuous-time signals, the CTFT is defined as follows:

$$X(f) = \int \left[x(t).\, e^{-j2\pi ft}\right] dt \tag{4.1}$$

For discrete-time signals, the DFT is defined as follows:

$$X(k) = \sum \left[x(n).\, e^{-j2\pi nk/N}\right] \tag{4.2}$$

In these equations, X(f) or X(k) represents the transformed signal in the frequency domain, and x(t) or x(n) represents the original signal in the time domain. 'f' represents the continuous frequency variable, 't' represents continuous time, 'k' represents the discrete frequency index, and 'n' represents the discrete time index. 'j' denotes the imaginary unit, and 'N' represents the length of the discrete-time signal.

- **Interpret the Frequency Domain Representation**: The transformed signal, X(f) or X(k), represents the frequency content of the original signal. It describes the amplitudes and phases of the various sinusoidal components present in the signal. The magnitude of X(f) or X(k) represents the amplitude, and the angle or phase represents the phase shift associated with each frequency component.
- **Obtain the Spectrum**: The magnitude of the transformed signal, |X(f)| or |X(k)|, is referred to signal's spectrum. The spectrum provides information about the strength or power of each frequency component present in the original signal. It helps identify dominant frequencies and analyse the frequency distribution.

➢ **Inverse Fourier Transform** (Optional): If desired, the transformed signal can be converted back to the time domain using the inverse Fourier transform. The inverse Fourier transform allows you to reconstruct the original signal from its frequency domain representation. The equations for the inverse transforms are the reverse of the Fourier transform equations. For continuous-time signals, the inverse CTFT is defined as follows:

$$x(t) = \int\left[X(f).e^{j2\pi ft}\right]df \tag{4.3}$$

For discrete-time signals, the inverse DFT is defined as follows:

$$x(n) = \frac{1}{N}\sum\left[X(k).e^{j2\pi nk/N}\right] \tag{4.4}$$

By applying the inverse Fourier transform, the signal can be restored to the time domain, allowing for further processing or analysis if required.

The Fourier transform is a powerful tool for frequency domain signal processing, enabling the analysis and manipulation of signals in terms of their frequency content. It forms the foundation for many advanced techniques and algorithms such as audio processing, telecommunications, image processing, and control systems.

4.2 FOURIER TRANSFORM

The mathematical relation between frequency domain signal processing and Fourier transformation is fundamental to understanding how Fourier transformation is applied in frequency domain signal processing. Frequency domain signal processing involves analysing and manipulating signals in the frequency domain, where the signals are represented as a function of frequency. The Fourier transformation, on the other hand, is the mathematical tool used to convert a signal from the time domain to the frequency domain. The Fourier transform (FT) and its variants, such as the Discrete Fourier Transform (DFT) or the Continuous Fourier Transform (CTFT), mathematically describe the relationship between a signal in the time domain and its frequency domain representation.

The Fourier transform essentially decomposes a signal into its constituent frequency components, providing information about the amplitudes and phases associated with each frequency. It represents the signal as a sum of complex sinusoidal functions with different frequencies. Once a signal is transformed into the frequency domain using the Fourier transform, various operations can be performed for analysis and processing. These operations involve manipulating the frequency domain representation of the signal, such as filtering, convolution, equalization, or spectral analysis. To perform these operations, the frequency domain representation of the signal is modified, and then the inverse Fourier transform is applied to obtain the processed signal in the time domain. The inverse Fourier transform is the mathematical operation that converts a signal from the frequency domain back to the time domain.

The inverse Fourier transform allows for reconstructing the original signal from its frequency domain representation, facilitating further processing or analysis in the time domain. In summary, the Fourier transformation provides the mathematical framework for converting a signal from the time domain to the frequency domain. It decomposes the signal into its constituent frequency components and enables frequency domain signal processing operations. After processing in the frequency domain, the inverse Fourier transform converts the signal back to the time domain.

4.2.1 Fast Fourier Transform

The mathematical relationship between the Fast Fourier Transform (FFT) and frequency domain signal processing is based on the principle that the FFT is an efficient algorithm for computing the Discrete Fourier Transform (DFT), which is a critical component of frequency domain signal processing. The FFT is a specific algorithm that efficiently computes the DFT of a sequence of samples. It exploits the properties of symmetry and periodicity in the DFT to reduce the required computations. The FFT algorithm enables fast calculation of the DFT, making it practical for real-time and large-scale signal processing applications. The relationship between the FFT and frequency domain signal processing can be described as follows:

a. Transforming a Signal to the Frequency Domain: The FFT converts a discrete-time signal from the time domain to the frequency domain. Given a sequence of N samples x[n], the FFT computes the DFT, representing the signal in terms of its constituent frequency components. The DFT of a discrete-time signal x[n] is given by:

$$X(k) = \sum\left[x(n).e^{-j2\pi nk/N}\right] \tag{4.5}$$

Here, X(k) represents the frequency domain representation of the signal, where k is the discrete frequency index, n is the discrete time index, and N is the length of the discrete-time signal. The FFT algorithm computes the DFT efficiently with a complexity of O(N log N), compared to the O(N^2) complexity of the direct computation of the DFT. This efficiency makes the FFT suitable for analyzing and processing signals in the frequency domain.

Frequency Domain Signal Processing Operations: Using the FFT, various operations can be performed on the frequency domain representation. Once a signal is transformed to the frequency domain using the FFT, various operations can be performed on the frequency domain representation. These operations involve manipulating the signal's frequency components to achieve desired processing goals.

b. Inverse FFT: After performing frequency domain signal processing operations on the frequency domain representation of a signal, the inverse FFT (IFFT) can be used to convert the processed signal back to the time domain. The IFFT algorithm is a variant of the FFT that computes the inverse DFT efficiently. It takes the frequency domain representation of the signal and computes the corresponding time-domain samples. The relationship between the FFT and frequency domain signal processing is closely tied to the ability to efficiently compute the DFT using the FFT algorithm. The FFT enables fast and efficient computation of the frequency domain representation of a signal, allowing for various operations and analysis in the frequency domain. The IFFT complements the FFT by facilitating the conversion of the processed signal back to the time domain for further analysis or use.

DFT presumes periodic signals. Any aperiodic signal with N points (x_0, x_1, x_2, ... x_{N-1}), will be considered as periodic with fundamental period T = N. Δt. The application of T for an aperiodic signal is shown in Fig. 4.2

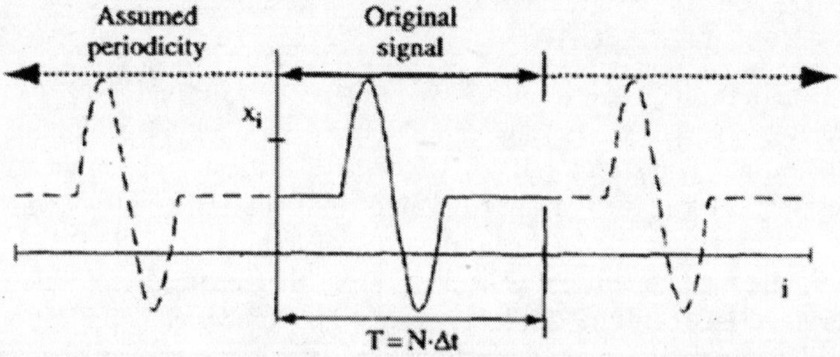

Fig 4.2: Pictorial view of Discrete Fourier transform (After Santamarina et al., 2005)

The highest frequency that can be resolved from a discrete time signal is 1/(2 Δt), called Nyquist frequency. So the Fourier series need not extend beyond N/2. Again, during transformation from time to frequency domain, the resolution will be symmetrical about f = 0. Since negative frequencies do not exist, hence, the DFT is usually defined from u = 0 to u = N/2. This is explained through Fig. 4.3. Following expression holds good :

$$f(t) = \sum_{u=-\alpha}^{\alpha} X_u \, e^{j\left(u\frac{2\pi}{T}t\right)} \underset{t_i = i.\Delta t}{\overset{T = N.\Delta t}{\rightarrow}} x_i = \sum_{u=-\alpha}^{\alpha} X_u \, e^{j\left(u\frac{2\pi}{N}i\right)} \quad (4.6)$$

where, $j = \sqrt{(-1)}$

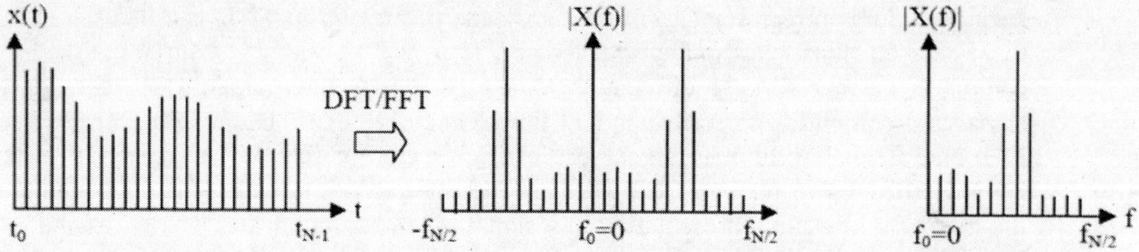

Fig 4.3: Transformation from time domain to frequency domain signal processing (After Santamarina et al., 2005)

The Fourier coefficient X_u is a complex number. Then, the amplitude and frequency of the component signal corresponds to X_u can be defined as :

$$\text{Amplitude, } \lceil X_u \rceil = \sqrt{(Re\, X_u)^2 + (Im\, X_u)^2}$$

$$\text{Frequency, } f = u.\frac{1}{T} = u.\frac{1}{\Delta t}$$

4.2.2 Implementation of FFT in MATLAB

In MATLAB, the function fft(x, N) is utilised to conduct the fast Fourier transform of a signal in the time domain. Following points should be remembered for the implementation procedure:

- X = fft(x, N) produces an array, containing 'N' complex Fourier coefficients. (default value of N = the number of sampling points in the time domain signal, x).
- If N > the sampling points in the time domain signal, the original signal will be extended by appending a certain number of zeroes to the end of the signal to include N points.
- Values of X are centred around index N/2, corresponding to $f_{N/2} = 1/(2\Delta t)$, not around zero. It is a two-sided definition of the spectrum.
- fft(x, N) will not transform the discrete-time data to a frequency corresponding to each element in X.
- Create a discrete time domain signal : $y = 2\sin\left(\frac{2\pi}{3t}\right)$

Let us take an example in Matlab. Suppose there are three periodic functions contaminated with a noise.

Say, sampling frequency f_s = 1.2 kHz

Sampling interval , $\Delta t = 1/f_s = 1/1200$ sec

First signal : a sine wave with amplitude 10 V and frequency 30 Hz

Second signal : a sine wave with amplitude 5 V and frequency 50 Hz

Third signal : a sine wave with amplitude 7 V and frequency 80 Hz

We have to see the time series of the combined input. The Matlab programme is discussed below:

The three signals are shown in Fig 4.4:

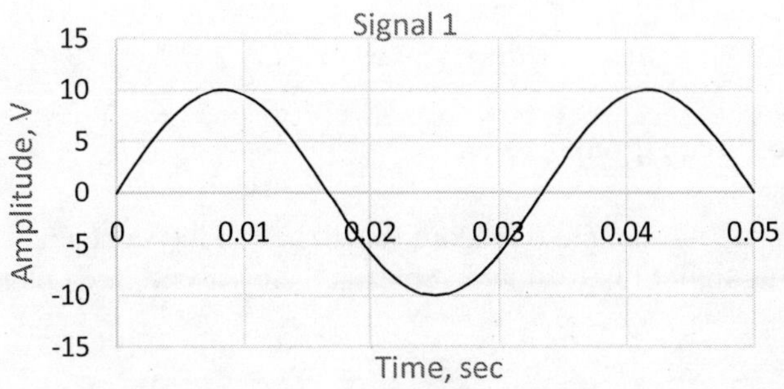

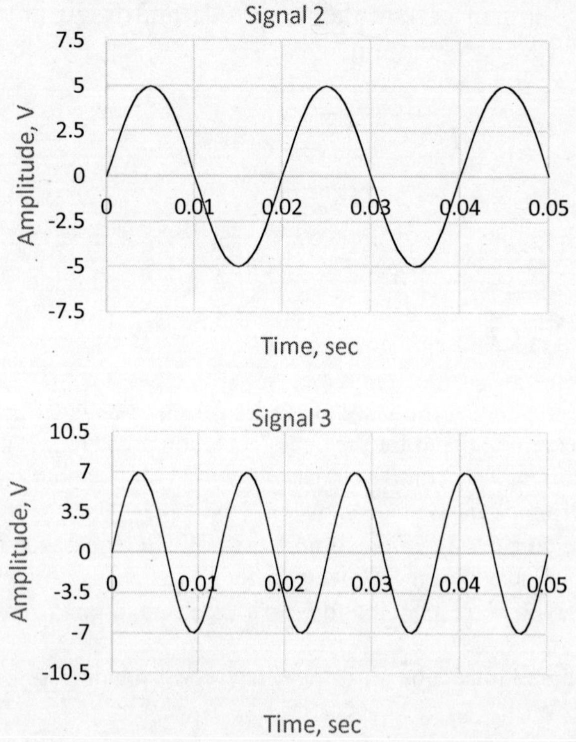

Fig 4.4: Three arbitrary signals

If all the three signals are combined, following figure, Fig 4.5 is obtained.

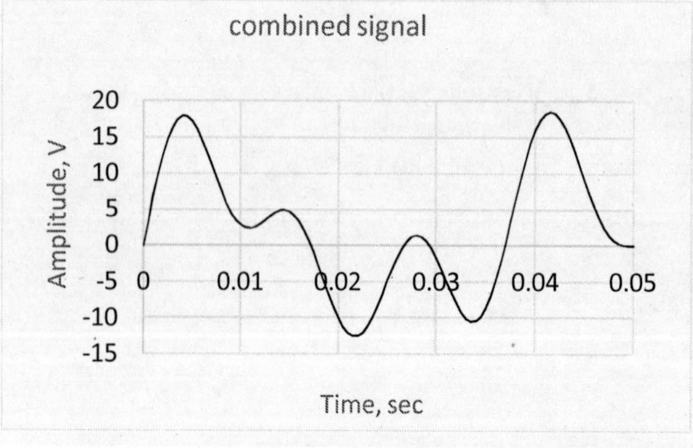

Fig 4.5: Combination of above three signals

If we add a noise to the above combined signal, we can put a random function with the same duration. The graph is plotted in Matlab vide subplot (3, 1, 1) which means there will be three plots in one column and the present graph is number 1. The title of the graph and the x and y labels are marked on the Matlab programme. The contaminated signal is marked as y. It can be seen from this figure that the signal does not clearly show the frequency contents. For obtaining the frequency contents, we have to carry out an FFT analysis.

The FFT of the contaminated signal for 256 (= 2^8) number of sampling points is obtained in the Matlab through the command 'Y = fft(y, 256). . The power spectral density (PSD) of the conjugate numbers in the FFT will give the frequency contents. It can be seen that frequency f is obtained for half of the total sampling points. The plot between PSD and the frequency contents is drawn in subplot 2 written as 'subplot(3,1,2)'. In case we want a zoom figure for the subplot to demarcate clearly the frequency contents, another subplot 3 is drawn with 40 sampling points. The frequency contents of the signal viz. 30 Hz, 50 Hz and 80 Hz are clearly obtained. The Matlab programme and the results are shown below.

```
close all;
clear all;
clc;
fs = 1200;   % sampling frequency
t = 0:1/fs:1   % time increment is 1/fs, starting from 0 to 1 sec
x = 10*sin(2*pi*30*t)+5*sin(2*pi*50*t)+7*sin(2*pi*80*t);
y = x + randn(size(t));
subplot(3,1,1);
plot(y(1:100));   % it gives a plot of three sine waves contaminated with noise
title('TIME SIGNAL CONTAINING NOISE');
xlabel ('time, t sec');
ylabel ('amplitude in V');
Y = fft(y,256);
PSD = Y.*conj(Y);
f = fs/256*(0:127);

subplot(3,1,2);
plot(f,PSD(1:128));
title('POWER SPECTRAL DENSITY');
xlabel ('frequency, f Hz');

subplot(3,1,3);
plot(f(1:40),PSD(1:40));
title('FREQUENCY CONTENTS');
xlabel ('frequency, f Hz');
```

The output of the above programme is show in Fig. 4.6

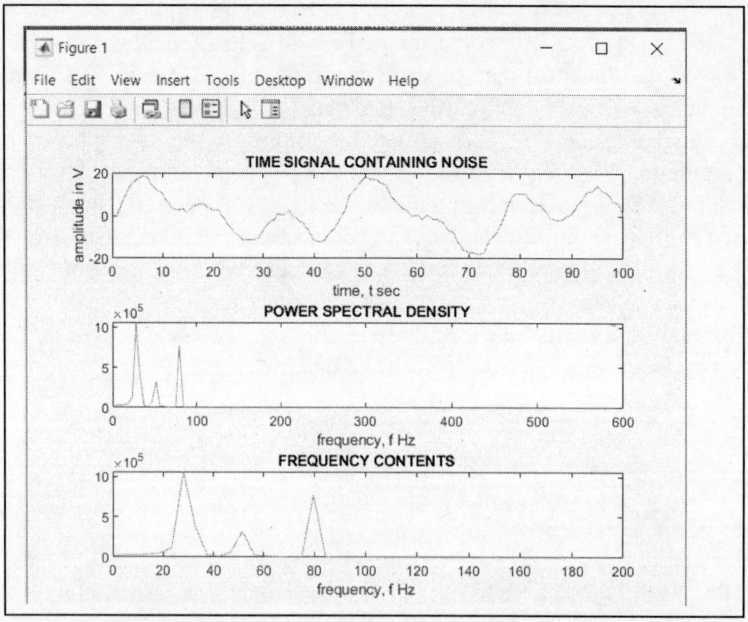

Fig 4.6: FFT analysis of a noise contaminated signal

Example 4.1

Ex. Retrieve the following two signals from a contaminated signal

$S1 = \sin 100\pi t$

$S2 = 2 \sin 240\ \pi t$

Take sampling frequency 1000 and signal duration 1.5 sec. Use Matlab .

Ans.

Initialize the Matlab programme with the data given. Obtain the combined signal without noise

```
close all;
clear all;
clc;
fs = 1000; % sampling frequency
T = 1/fs;  %sampling period
L = 1500;  % length of the signal 1/fs*1.5
t = (0:L-1)*T % time increment
x = sin(100*pi*t)+ 2*sin(240*pi*t);
subplot(4,1,1);
plot(x(1:100)); % it gives a plot of two sine waves upto a time 100 sec
title('two signals');
xlabel ('time, t sec');
ylabel ('amplitude ')
```

The signal is shown in the first graph of Fig.4.7

The combined signal is then contaminated with a zero mean white noise with variance 4. The contaminated signal is shown in the second graph of Fig.4.7. It can be seen that it is very difficult to locate the original signals from this contaminated signal.

```
y = x + 2*randn(size(t));
subplot(4,1,2);
plot(1000*t(1:50), y(1:50)); % it gives a plot of two sine waves contaminated
with noise
title('CONTAMNATED TIME SIGNAL');
xlabel ('time, t millisec');
ylabel (' y(t)');
```

Compute the Fourier transform of the signal y. Then, compute the two-sided spectrum SP2. Next, compute the single-sided spectrum SP1 based on SP2 and even-valued signal length L.

```
Y = fft(y);
SP2= abs(Y/L);
SP1 = SP2(1:L/2+1);
SP1(2:end-1)= 2*SP1(2:end-1);
```

Define the frequency domain f and plot the single -sided amplitude spectrum SP1. The amplitudes are nearly 1 and 2 because of noise contamination. The frequency plot is shown in subplot 3 in Fig 4.7

```
f = fs*(0:(L/2))/L;
subplot(4,1,3);
plot(f,SP1);
title('Single-Sided Amplitude Spectrum of y(t)');
xlabel('f (Hz)');
ylabel('|SP1(f)|');
```

In order to clearly get the original signal frequency 50 Hz and 120 Hz the subplot 3 may be zoomed as shown in subplot 4.

```
subplot(4,1,4);
plot(f(1:200),SP1(1:200));
title('FREQUENCY CONTENTS');
xlabel ('frequency, f Hz');
```

The output of the Matlab file is shown in Fig 4.7

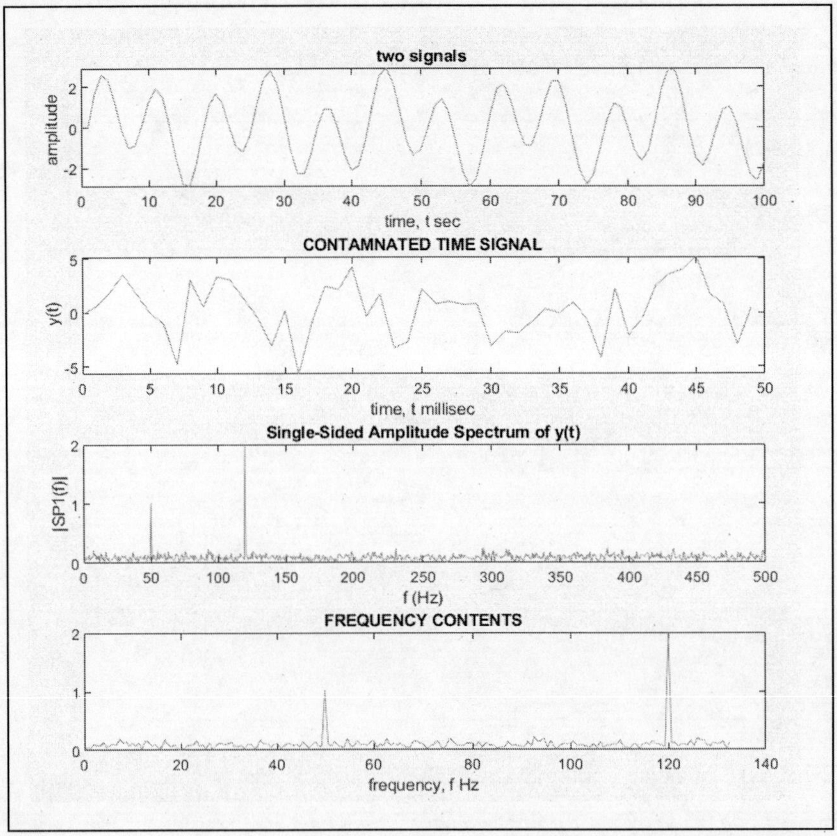

Fig 4.7: FFT of two sine waves contaminated with noise

The non-contaminated signals can be obtained by taking the FFT of x instead of y. Following modifications may be made

```
Y = fft(x);
SP2= abs(Y/L);
SP1 = SP2(1:L/2+1);
SP1(2:end-1)= 2*SP1(2:end-1);
%
f = fs*(0:(L/2))/L;
subplot(4,1,3);
plot(f,SP1);
title('Single-Sided Amplitude Spectrum of y(t)');
xlabel('f (Hz)');
ylabel('|SP1(f)|');
```

The original frequencies in the combined signal are shown below (Fig 4.8)

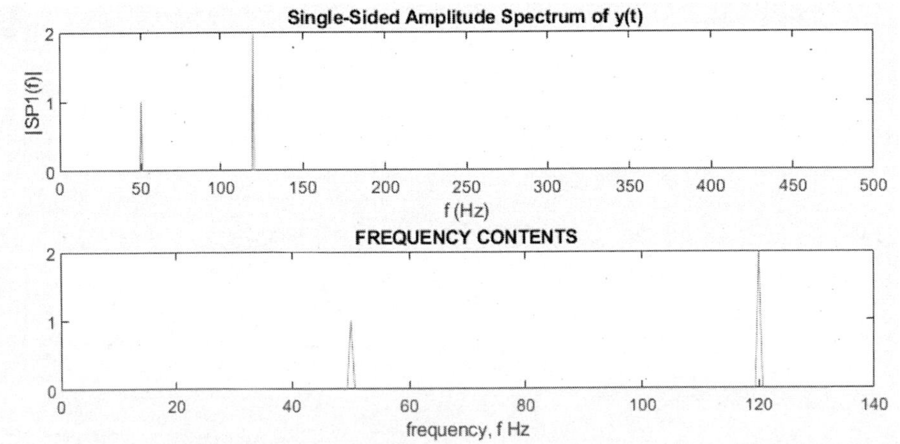

Fig 4.8: FFT of two sine waves without contamination

Example 4.2

Ex. Obtain the FFT of the following vector. Show that the IFFT gives back the original vector. Use Matlab.

xx = [2 4 7 8 9 10]

It is seen earlier that the Matlab has FFT code. It also has the IFFT code. Following programme may be written in Matlab.

```
clear all;
close all;
clc;
xx = [2 4 7 8 9 10];
yy = fft(xx) % it will give the FFT with six complex values
zz = ifft(yy) % it will give back the original vector
```

Following output is obtained :

yy =
40.0000 + 0.0000i -7.0000 + 6.9282i -5.0000 + 3.4641i -4.0000 + 0.0000i -5.0000 - 3.4641i
-7.0000 - 6.9282i

zz = 2 4 7 8 9 10 (original signal)

4.2.3 Implementation of FFT in Mathcad

In Mathcad, the function fft(x) is used to achieve the fast Fourier transform of a signal in the time domain. The following steps are to be followed for the implementation procedure.

- X = fft(x) needs the input signal in the time domain with 2n data points, and will return an array of contains $2^{n-1}+1$ complex Fourier coefficients, corresponding to the frequency from 0 to $f_{N/2}= 1/(2\Delta t)$. It is a one-sided definition of the spectrum

- fft(x) will not transform the discrete-time data to frequency corresponding to each element in X.

- If the number of sampling points in the input signal is not equal to 2^n, the user needs to manually extend the signal by appending a certain number of zeros to the end of the signal to make it include 2^n points, before using the fft function.

Note: the cfft(x) function can be applied to a signal of any length, but it could be slow and not intuitive. Its use is discouraged.

 # Example: Frequency domain analysis of thermistor readings

Temperature readings taken at Lake Raleigh Dam during 06/2011 to 08/2011 are shown below in both time domain and frequency domain (Fig 4.9).

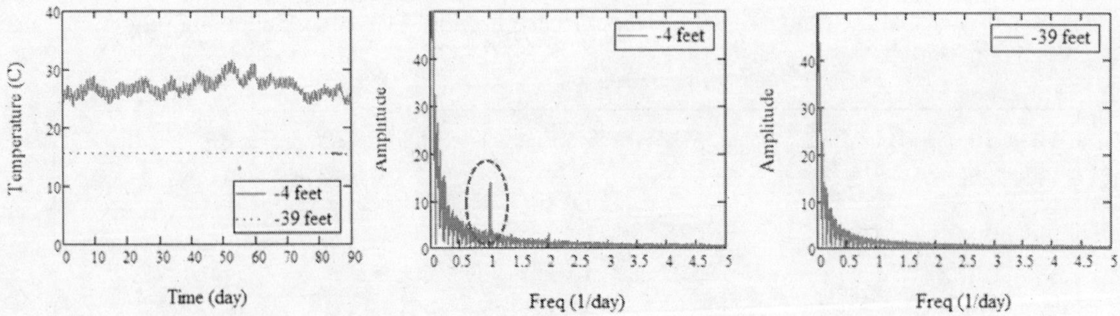

Fig 4.9: Graphical representation of time domain and frequency domain analysis (After Santamarina et al., 2005).

In the time domain, it can be seen that the temperature near the ground surface fluctuated while the temperature at 39 ft. deep was fairly stable. Frequency domain analysis helps further reveal that the near ground surface temperature was subjected to a daily cyclic change, which is indicated by a peak at f = 1 day^{-1}.

4.3 FILTERS IN THE FREQUENCY DOMAIN

Noise Reduction Filters

Noise reduction filters in frequency domain signal processing and analysis refer to techniques that aim to attenuate or remove unwanted noise components from a signal by manipulating its frequency domain representation. These filters are designed to selectively suppress noise frequencies while preserving the desired signal components. Several commonly used noise reduction filters in the frequency domain include:

a. **Low-Pass Filters**: Low-pass filters are designed to attenuate high-frequency components while preserving the low-frequency components of a signal. In noise reduction applications, a low-pass filter in the frequency domain can suppress high-frequency noise. By setting an appropriate cutoff frequency, the filter allows only the low-frequency components, including the desired signal, to pass through, effectively reducing the noise.
b. **High-Pass Filters**: High-pass filters attenuate low-frequency components while allowing high-frequency components to pass through. In the context of noise reduction, a high-pass filter can be applied in the frequency domain to suppress low-frequency noise. The filter blocks frequencies below a specific cutoff, allowing only the high-frequency components, including the desired signal, to remain.
c. **Band-Pass Filters**: Band-pass filters selectively allow a specific frequency range, or band, to pass through while attenuating frequencies outside that range. In noise reduction, a band-pass filter in the frequency domain can be utilized to target specific noise bands. By setting appropriate lower and upper cutoff frequencies, the filter attenuates frequencies outside the desired band, effectively reducing the noise within that band.
d. **Notch Filters**: Notch filters, also known as band-stop filters, are designed to suppress a narrow frequency range, typically centered around a specific frequency. They attenuate a specific frequency band while allowing other frequencies to pass through. Notch filters are particularly useful for reducing noise caused by specific interfering frequencies or harmonics. By specifying the centre frequency and the bandwidth of the notch, the filter can selectively attenuate the noise components in that range.
e. **Wiener Filters**: Wiener filters are adaptive filters commonly used for noise reduction applications. They aim to minimize the mean square error between the desired signal and the filtered signal. The noise Wiener filter estimates the power spectral density (PSD) of the desired signal and noise in the frequency domain. By taking the ratio of the desired signal's PSD to the sum of the desired signal and noise PSD, the filter computes the Wiener gain, representing the attenuation amount applied to different frequency components. The Wiener filter adaptively adjusts its frequency response based on the signal and noise statistics, making it effective in reducing noise while preserving the desired signal.

These noise reduction filters can be applied in various domains, including audio processing, image processing, telecommunications, and biomedical signal processing. They are implemented by manipulating the frequency domain representation of the signal, such as multiplying the frequency spectrum by the desired filter response. The inverse Fourier transform is then applied to convert the filtered signal back to the time domain for further processing or analysis. It's worth noting that the selection and design of an appropriate noise reduction filter depend on the specific characteristics of the noise and the desired signal. Different noise sources may require different filter configurations to achieve optimal noise reduction while minimizing the impact on the desired signal.

Implementation Procedure

- Convert signal from time domain to frequency domain by FFT: $\underline{X}$=FFT($\underline{x}$).
- Plot Magnitude $|X_u|$ vs. f_u to determine the frequency of interest
- Choose the type of filter, select filter parameters, define filter array $\boldsymbol{G}$ based on frequency $\boldsymbol{f}$
- Apply filter to the signal's amplitude in the time domain; i.e., multiply point by point:

$$|Y_u| = G_u . |X_u|$$

- Compute the filtered signal: $= |Y_u|[\cos(\varphi_u) + i \sin \varphi_u]$; where φ is phase angle.
- Convert the signal back in the time domain by inverse FFT: y=IFFT(Y)

 # Example: Removing noise from inclinometer readings with filter

Fig 4.10 shows an example of use of filter in removing the noise in a signal. Original signal is shown in Fig 4.10 (a) over a period of 90 days. Unknown noise is presented in the reading of an inclinometer at different depths. FFT analysis is shown in Fig 4.10 (b) for the reading at 11 feet depth. A low pass filter with cut-off frequency = 1/day is applied. Converting frequency domain analysis to time domain again, the noise free signals are obtained.

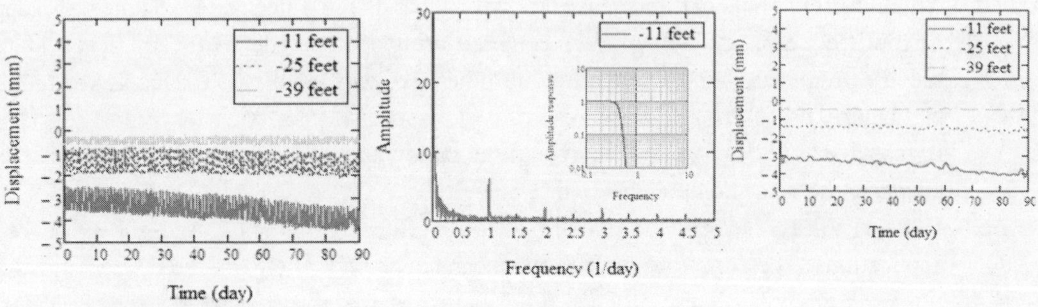

(a) Original signals in the time domain

(b) A signal in the frequency domain and the low-pass filter applied

(c) Signals in the time domain after the application of filter

Fig 4.10: Graph showing time domain and frequency domain analysis without and with filter (After Wang, 2009).

4.3.1 Use of filters in Matlab

Some filter commands are available in the Matlab. If the following command is written

```
y = filter (b, a, x)
```

The command filters the input data x using a rational transfer function defined by the numerator and denominator coefficients b and a.

If x is a vector, then the filter returns the filtered data as a vector of the same size as x.

If x is a vector, then filter returns the filtered data as a vector of the same size as x.

If a(1) is not equal to 1, then filter normalizes the filter coefficients by a(1), therefore, a(1) must be non-zero.

The simplest filter is moving-average filter. The moving-average filter slides a window of length *windowsize* along the data, computing averages of the data contained in each window. The following difference equation defines a moving-average filter of a vector x

$$y(n) = \frac{1}{windowsize}(x(n) + x(n-1) + \cdots + x(n-(windowsize-1))) \tag{4.7}$$

Example 4.3

Ex. Obtain the filtered data of a sine wave corrupted by a random noise. Use Matlab.

Adopt a window size 10.

It is expected that the windowsize 10 is quite large, and the corrupted sine wave will be smoothly filtered. The Matlab programme is as follow:

```
clc;
clear all;
close all;
t= linspace(-pi,pi,100); % sine wave (-pi to pi) with 100 vectors
rng default  %initialize random number generator
x = sin(t) + 0.25*rand(size(t)); % sine wave is corrupted here
windowsize = 10;
b = (1/windowsize)*ones(1,windowsize);
a  = 1;
y = filter (b,a,x);
plot(t,x);
hold on;
plot(t,y);
legend('Input data', 'Filtered data');
```

Following is the output of the programme:

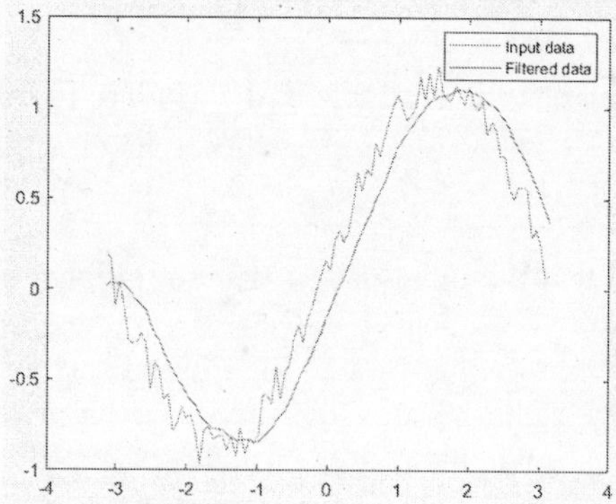

Ex. A traffic data shows number of vehicles on a road over 24 hours. Obtain the filtered data using a window size 4.

Time in hours :	1	2	3	4	5	6	7	8	9	10	11	12
Number of vehicles :	10	9	11	8	45	30	75	60	50	40	15	20

Time in hours :	13	14	15	16	17	18	19	20	21	22	23	24
Number of vehicles :	22	23	30	35	40	80	45	30	10	12	10	5

Ans

consider a = 1

since windowsize = 4, hence b = [¼ ¼ ¼ ¼]

The following difference equation describes a filter that averages time-dependent data with respect to the current hour and the three previous hours of data :

$$y(n) = \frac{1}{4}\,x(n) + \frac{1}{4}\,x(n-1) + \frac{1}{4}\,x(n-2) + \frac{1}{4}\,x(n-3)$$

The Matlab programme is shown below :

```
clc;
clear all;
close all;
t= 1:24;
x = [10 9 11 8 45 30 75 60 50 40 15 20 22 23 30 35 40 80 45 30 10 12 10 5]';
windowsize = 4;
b = (1/windowsize)*ones(1,windowsize);
a  = 1;
y = filter (b,a,x);
plot(t,x,'--',t,y,'-');
legend('Input data', 'Filtered data');
```

Example 4.4

The output of the Matlab programme is shown below :

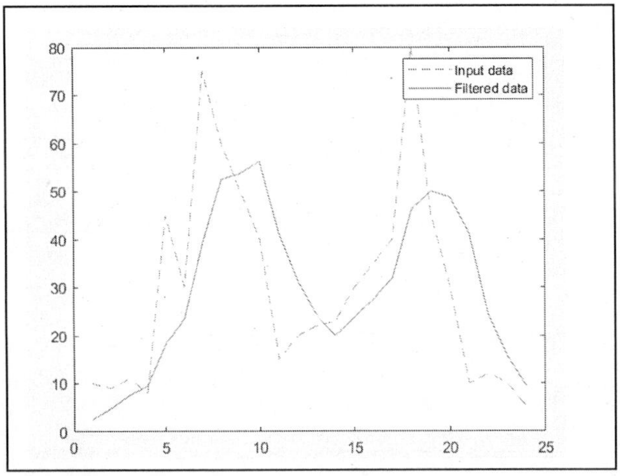

4.4 LEAKAGE

Leakage is a phenomenon that occurs in frequency domain signal processing and analysis when the frequency components of a signal spread or "leak" into neighbouring frequency bins, leading to inaccurate frequency representations. It results from using finite-length time windows or discrete sampling in the time domain.

Leakage can be better understood by considering the Discrete Fourier Transform (DFT) as an example. The DFT computes the frequency domain representation of a discrete-time signal by dividing the time domain signal into equally spaced segments (time windows) and calculating the Fourier transform of each segment.

When a time window is applied to a signal, it effectively assumes that the signal within that window is periodic. However, if the signal is not perfectly periodic within the window, the Fourier transform of the windowed segment exhibits spectral leakage. This means that the frequency components of the signal "leak" or spread across adjacent frequency bins, resulting in inaccurate representation of the signal's frequency content. Consider a periodic signal as shown in Fig 4.11. Now suppose our observation time is also periodic and equals to the period of the signal as shown in the Fig 4.11. If the observation time is repeated, the original signal is obtained. But if the observation time is aperiodic as shown in fig Fig 4.11, then repetition of the observed signal will show a discontinuity as shown in Fig 4.11. The sharp discontinuity gives rise to leakage.

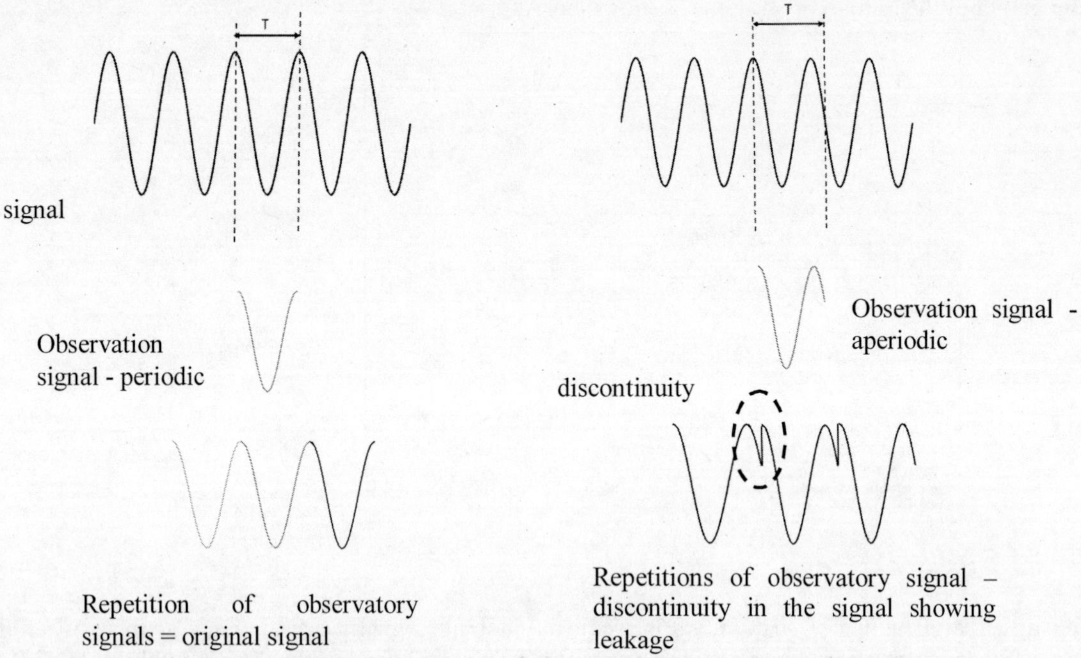

Fig. 4.11: Repetitions of periodic and aperiodic signals

Leakage is caused by the discontinuities at the beginning and end of the time window, as well as the mismatch between the frequency of the signal and the discrete frequency bins of the DFT. These discontinuities introduce high-frequency components that do not exist in the original signal, leading to spectral spreading and distortion.

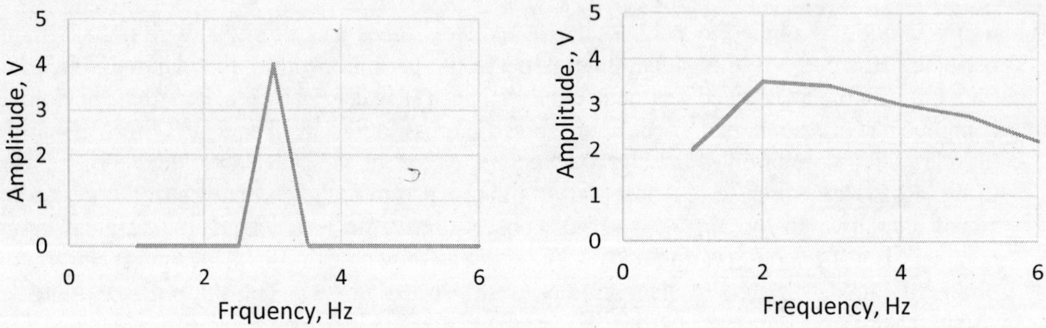

(a) Spectral line of 4V sine wave with Frequency 3 Hz

(b) Spectral line of 4V sine wave with with frequency 2.5 Hz

Fig 4.12: Leakage for non-integer frequency of periodic signal

The leakage effect is particularly prominent when the signal frequency does not align perfectly with the discrete frequency bins of the DFT. If the signal frequency is not an integer multiple of the frequency bin spacing (frequency resolution) of the DFT, the leakage effect becomes more significant. Consider a 4V sine wave signal with frequency 3Hz. The spectral line is as shown in Fig. 4.12. If a 4V sine wave with frequency 2.5 Hz is chosen, the spectral line will not show 2.5 Hz, but it will spread over other frequencies. This is called leakage. The spectral energy leaks across all spectral lines in unpredictable manner.

To mitigate leakage, several techniques can be employed:

1. **Windowing**: Applying a window function to the signal before the Fourier transform, can reduce the discontinuities at the edges of the time window. Window functions, such as the Hamming, Hanning, or Blackman windows, taper the signal gradually to zero at the edges, reducing spectral leakage. However, windowing introduces a trade-off between reducing leakage and widening the main lobe of the frequency response. Fig 4.13 (a) shows an aperiodic signal. If this signal is multiplied with a Hanning window shown in Fig 4.13 (b) with unit amplitude and range zero to zero, the modified signal is shown in Fig 4.13 (c). Repetitions of this modified signal is shown in Fig 4.13 (d). It can be seen that the discontinuity or sharp edges have been removed. Thus, leakage is reduced with window function.

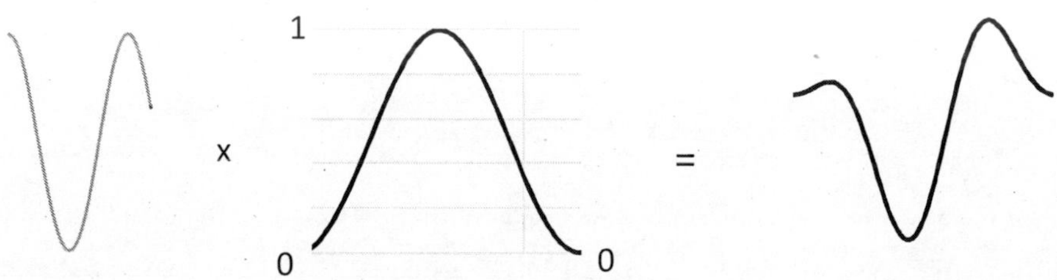

(a) Aperiodic signal (b) Hanning window function 0 to 0 and amplitude = 1 (c) Modified signal

$A = \sin^2(i\,\pi t/N)$, where i = 0, 1, 2, … N-1

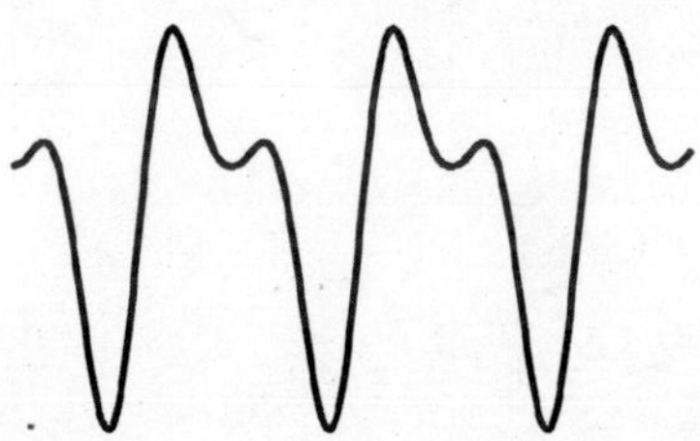

(d) Repetitions of modified signal

Fig 4.13: Removal of leakage using window function

2. **Zero-padding**: Increasing the length of the time window by adding zeros to the signal can improve frequency resolution and reduce spectral leakage. Zero-padding effectively interpolates additional frequency bins, providing a more accurate representation of the signal's frequency content. However, zero-padding does not completely eliminate leakage; it only reduces its impact.

FFT generally requires the number of data points N to be an integer power of 2 (e.g., 1024, 2048, 4096, etc.) to attain maximum computational efficiency. If not, zeroes can be added at the end of the original record. This is called zero padding (Fig 4.14).

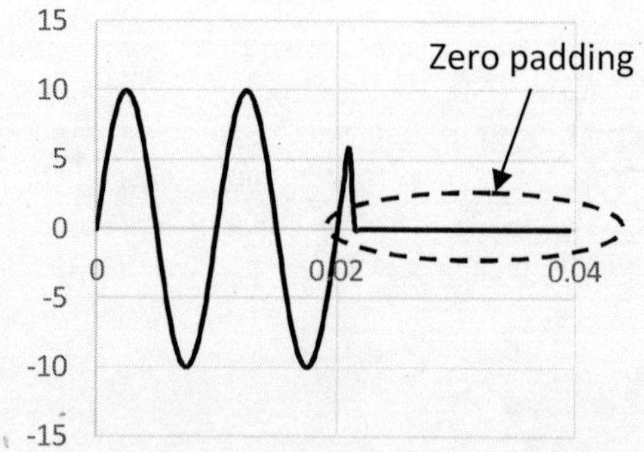

Fig 4.14: Pictorial view of zero padding zone.

3. **Window Design**: Selecting an appropriate window function can help minimize spectral leakage for specific applications. Different window functions have different trade-offs between leakage reduction and frequency resolution. Choosing the optimal window function depends on the specific requirements of the application.
4. **Window Overlap**: Overlapping consecutive time windows can improve frequency resolution and reduce leakage. By overlapping segments of the signal and computing the Fourier transform, the spectral leakage is averaged out over multiple windows, resulting in a smoother frequency representation.
5. **Advanced Techniques**: More advanced techniques, such as adaptive algorithms and signal modelling, can be employed to estimate and compensate for leakage effects. These techniques involve estimating the spectral leakage based on the characteristics of the window and applying corrections to the frequency domain representation.

It is important to note that while these techniques can mitigate the effects of leakage, they do not eliminate it completely. Leakage is an inherent limitation of finite-length time windows and discrete frequency sampling. Understanding and managing leakage is essential in ensuring accurate frequency domain analysis and interpretation of signals.

4.5 FREQUENCY RESOLUTION

Frequency resolution refers to the ability to distinguish between closely spaced frequency components in the frequency domain. It is a critical parameter in frequency domain signal processing and analysis, as it determines the level of detail that can be resolved in the frequency representation of a signal.

The frequency resolution is directly related to the length or duration of the time window used in the analysis. In general, a longer time window provides a better frequency resolution, allowing for the identification and separation of closely spaced frequency components. Conversely, a shorter time window leads to poorer frequency resolution, making it more difficult to distinguish between nearby frequencies.

The relationship between the time window length (T) and the frequency resolution (Δf) is determined by the uncertainty principle, which states a fundamental trade-off between time and frequency resolution. The uncertainty principle implies that the product of the time duration and frequency resolution is bounded:

$$\Delta f . T \geq 1 \quad (4.8)$$

The sampling duration, T can be expressed as $(\Delta t. number\ of\ sampling\ points)$

This means that as the time window length increases (higher T), the frequency resolution improves (smaller Δf). Similarly, a shorter time window (lower T) leads to broader frequency resolution (larger Δf).

The frequency resolution can also be related to the sampling rate (Fs) of the signal. The sampling rate determines the spacing between discrete frequency bins in the frequency domain. The frequency resolution (Δf) is inversely proportional to the length of the signal in the time domain (N) :

$$\Delta f = F_s / N \quad (4.9)$$

This relation indicates that increasing the sampling rate or the length of the signal improves the frequency resolution. It is important to note that frequency resolution differs from spectral leakage and spectral bin width, which are related phenomena in frequency domain analysis. Spectral leakage refers to the spreading of spectral energy of a signal across adjacent frequency bins due to finite-length time windows or discrete sampling. Spectral leakage can affect frequency resolution, as it may cause overlapping or smearing of frequency components. Techniques such as windowing and zero-padding can help mitigate the effects of spectral leakage. Spectral bin width refers to the width or spacing between adjacent frequency bins in the frequency domain. The bin width is determined by the total length of the signal and the sampling rate. Smaller bin widths allow for finer frequency resolution and better separation of frequency components. In summary, frequency resolution in frequency domain signal processing and analysis relates to the ability to distinguish between closely spaced frequency components. It is determined by the length of the time window or the total length of the signal, as well as the sampling rate. Longer time windows and higher sampling rates result in better frequency resolution, allowing for finer distinctions between frequency components. Spectral leakage and spectral bin width are additional factors that can influence the practical frequency resolution achieved in a given analysis.

An example may be cited to explain the frequency resolution. Suppose, in a time domain there are 8192 sampling points and time between consecutive sampling points (Δt) = 1/48000. The signal contains two component frequencies f_1 = 1000 Hz and f_2 = 1003 Hz. We want to make frequency resolution.

From equation (4.1), we get $\Delta f = \frac{48000}{8192} = 5.86\ Hz$

In the present case Δf = 1003 – 1000 = 3 Hz.

Thus, the FFT analysis will not differentiate the two frequencies. In order to differentiate the frequencies, the FFT size may be increased from 8192 to 32768 using zero padding. But the real FFT frequency resolution will not change from 5.86 Hz. In such a case, the frequency resolution can be reduced by:

a) Increasing the number of sampling points, or
b) Increasing the value of Δt, i.e., reducing the sampling frequency.

In the above problem, if sampling points are increased to 32768, the frequency resolution is given by

$$\Delta f = \frac{48000}{32768} = 1.46\ Hz < 3\ \text{Hz}$$

The FFT analysis will clearly show two peaks, one at 1000 Hz and the other at 1003 Hz.

UNIT SUMMARY

What have we learnt from module 4?

Module 4 is about the frequency domain data analysis and in-depth understanding of physical phenomena behind monitoring data. Following topics have been covered in this module :

1. Basic definitions of time domain and frequency domain analysis.
2. Importance of the frequency domain analysis and the corresponding principles.
3. Brief discussion on Fourier Transform and Fast Fourier Transform.
4. How the time domain analysis can be converted to frequency domain analysis.
5. Basic idea about the noise reduction with filters and its application
6. Concepts of Leakage, and Frequency resolution.

MULTIPLE CHOICE QUESTIONS

Q.1. Which type of signal facilitates easier noise removal ?
  a) Analog Signal
  b) Digital Signal
  c) Random Signal
  d) Sine Signal

Q.2. Low pass filters remove the sharpness of an image.
  a) True
  b) False

Q.3. Which filter should be used to remove impulse noises?
  a) Mean filter
  b) Gaussian filter
  c) Median filter
  d) Weiner filter

Q.4. The characteristic of windowing the signal called "Leakage" is the power leaking into the entire frequency range.

a) True
b) False

Q.5. Which of the following is the advantage of Hanning window over rectangular window?

a) Less side lobes
b) More side lobes
c) More width of the main lobe
d) None of the mentioned

Q.6. If the signal to be analyzed is an analog signal, we would pass it through an anti-aliasing filter with B as the bandwidth of the filtered signal and then the signal is sampled at a rate __________.

a) $F_s \leq 2B$
b) $F_s \leq B$
c) $F_s = 2B$
d) $F_s \geq 2B$

Q.7. What is the highest frequency contained in the sampled signal?

a) Fs/2
b) 2Fs
c) Fs
d) None of thc mcntioned

Q.8. Which of the following is true regarding the number of computations required to compute an N-point DFT?

a) N^2 complex additions and N(N-1) complex multiplications
b) N^2 complex multiplications and N(N-1) complex additions
c) N^2 complex multiplications and N(N+1) complex additions
d) N^2 complex additions and N(N+1) complex multiplications

Q.9. If X(k) is the N/2 point DFT of the sequence x(n), then what is the value of X(k+N/2) ?

a) F1(k)+F2(k)
b) F1(k)-WNk F2(k)
c) F1(k)+WNk F2(k)
d) None of the mentioned

Q.10. For a decimation-in-time FFT algorithm, which of the following is true?

a) Both input and output are in order

b) Both input and output are shuffled

c) Input is shuffled and output is in order

d) Input is in order and output is shuffled

Q.11. What is the main function of (A/D) or ADC converter?

a) Converts Analog to Digital signal

b) Converts Digital to Analog Signal

c) All of the mentioned

d) None of the mentioned

Q.12. Which of the following techniques is commonly used to convert a signal from the time domain to the frequency domain?

a) Fourier Transform

b) Laplace Transform

c) Z-Transform

d) Wavelet Transform

Q.13. In frequency domain analysis, the magnitude of a signal represents:

a) Amplitude

b) Phase

c) Frequency

d) Energy

Q.14. What does the phase spectrum of a signal in the frequency domain indicate?

a) The time at which different frequencies occur

b) The frequency content of the signal

c) The amplitude of each frequency component

d) The time delay associated with each frequency component

Q.15. Which of the following is NOT a common application of frequency domain signal processing?

a) Filtering

b) Compression

c) Noise reduction

d) Edge detection

Q.16. The process of converting a signal from the frequency domain back to the time domain is achieved through:

a) Fourier Transform

b) Inverse Fourier Transform

c) Wavelet Transform

d) Laplace Transform

Q.17. A signal with a sharp peak in the frequency domain indicates:

a) High frequency content

b) Low frequency content

c) Narrow bandwidth

d) Wide bandwidth

Q.18. Which of the following techniques is commonly used to remove unwanted frequency components from a signal?

a) High-pass filtering

b) Low-pass filtering

c) Band-pass filtering

d) Notch filtering

Q.19. The frequency resolution of a signal in the frequency domain is determined by:

a) The sampling rate

b) The duration of the signal

c) The number of frequency bins

d) The amplitude of the signal

Q.20. What is the main advantage of using frequency domain analysis over time domain analysis?

a) It provides information about the energy distribution of the signal.

b) It simplifies complex signals into their constituent frequencies.

c) It reduces computational complexity.

d) It is less sensitive to noise.

Q.21. Which of the following statements about the Fourier Transform is true?

a) It converts a signal from the frequency to the time domain.

b) It is only applicable to periodic signals.

c) It decomposes a signal into its frequency components.

d) It is limited to analyzing signals with a fixed duration.

Q.22. The term "spectral leakage" refers to:

a) Loss of signal energy during Fourier Transform.

b) Spreading of signal energy into adjacent frequency bins.

c) Sharp peaks in the frequency spectrum.

d) Attenuation of low-frequency components.

Q.23. Which windowing function is commonly used to mitigate spectral leakage in frequency analysis?

a) Rectangular window

b) Hamming window

c) Blackman window

d) Kaiser window

Q.24. In frequency domain signal processing, the term "Nyquist frequency" refers to:

a) The highest frequency component present in the signal.

b) Twice the highest frequency component present in the signal.

c) Half the sampling frequency.

d) The frequency at which aliasing occurs.

Q.25. Which of the following statements about the power spectrum of a signal is true?

a) It represents the amplitude of each frequency component.

b) It represents the energy distribution across different frequencies.

c) It is identical to the magnitude spectrum.

d) It is calculated using the phase spectrum.

Q.26. The process of convolving a signal with a filter kernel in the frequency domain is known as:

a) Filtering

b) Modulation

c) Deconvolution

d) Convolution

Answers of Multiple Choice Questions
1(b), 2(a), 3(c), 4(a), 5(a), 6(d), 7(a), 8(b), 9(b), 10(d), 11(a), 12(a), 13(a) 14(d), 15(d), 16(b), 17(c), 18(b), 19(a), 20(b), 21(c), 22(b), 23(c), 24(c), 25(b), 26(d)

SHORT AND LONG ANSWER TYPE QUESTIONS

Q.1. Define frequency domain analysis.

Q.2. Differentiate between time and frequency domain analysis.

Q.3. How time domain frequency can be transformed into frequency domain analysis?

Q.4. How FT can be differentiated from FFT?

Q.5. What is DFFT?

Q.6. Write short notes on (a) Zero padding

(b) Spectrum analysis

(c) Filtering

(d) Convolution

(e) Equalization

(f) Windowing

Q.7. How FFT can be implemented in MATLAB?

Q.8. Define leakage. How can leakage be mitigated?

Q.9. In which field are frequency domain analysis and signal processing used?

Q.10. Write a short note on frequency resolution.

EXERCISES

Ex. A signal $x(n)$ has the following Fourier transform:

$$X(w) = \frac{1}{1 - ae^{-jw}}$$

Determine the Fourier transforms of the following signals:

(a) $x(2n+1)$

(b) $e^{\pi n/2}x(n+2)$

(c) $x(-2n)$

Ans:

(a)

$$X_1(w) = \sum_n x(2n+1)e^{-jwn}$$
$$= \sum_k x(k)e^{-jwk/2}e^{jw/2}$$
$$= X\left(\frac{w}{2}\right)e^{jw/2}$$
$$= \frac{e^{jw/2}}{1 - ae^{jw/2}}$$

(b)

$$X_2(w) = \sum_n x(n+2)e^{\pi n/2}e^{-jwn}$$
$$= -\sum_k x(k)e^{-jk(w+\frac{j\pi}{2})}e^{j2w}$$
$$= -X(w + j\pi/2)e^{j2w}$$

(c)

$$X_3(w) = \sum_n x(-2n)e^{-jwn}$$
$$= -\sum_k x(k)e^{-jwk/2}$$
$$= X\left(-\frac{w}{2}\right)$$

Example 4.5

Ex.

Given a discrete-time signal $[n]=\{1,2,3,4\}$ find its Discrete Fourier Transform (DFT) $X[k]$.

Ans:

$$\mathrm{X[k]} = \sum_{n=0}^{N-1} x[n].\, e^{-j2\pi kn/N}$$

Substituting the values:

$X[0] = 1\cdot e^{-j0} + 2\cdot e^{-j0} + 3\cdot e{-j0} + 4\cdot e^{-j0}$

$= 1+2+3+4 = 10$

$X[1] = 1\cdot e^{-j2\pi/4} + 2\cdot e^{-j2\pi/4} + 3\cdot e^{-j6\pi/4} + 4\cdot e^{-j8\pi/4}$

$= 1-2j-3-4j = -2-6j$

$X[2] = 1\cdot e^{-j4\pi/4} + 2\cdot e^{-j8\pi/4} + 3\cdot e^{-j12\pi/4} + 4\cdot e^{-j16\pi/4}$

$= 1-3-5+7 = 0$

$X[3] = 1\cdot e^{-j6\pi/4} + 2\cdot e^{-j12\pi/4} + 3\cdot e^{-j18\pi/4} + 4\cdot e^{-j24\pi/4}$

$= 1+4j-3-4j = -2$

So, $\mathrm{X[k]} = \{10, -2-6j, 0, -2\}$.

Example 4.6

Example 4.7. A continuous-time signal $(t) = \sin(2\pi f_0 t)\mathrm{x(t)}$ is sampled at a frequency of $f_s = 100$ Hz. What is the minimum sampling frequency required to avoid aliasing if the signal frequency $f_0 = 40$ Hz?

Answer:

To avoid aliasing, according to the Nyquist theorem, the sampling frequency (f_s) must be at least twice the maximum frequency component in the signal (f_{max}). In this case, the maximum frequency component is the signal frequency $f_0 = 40$ Hz.

So, the minimum sampling frequency required to avoid aliasing (f_{min}) is given by:

$f_{min} = 2\times f_0$

Substituting the given value $f_0 = 40$ Hz:

$f_{min} = 2\times 40 = 80$ Hz

Therefore, the minimum sampling frequency required to avoid aliasing is

$f_{min} = 80$ Hz.

Example 4.8. Find the inverse Discrete Fourier Transform (IDFT) of $[k]$={3, 1+j, −2, 1−j}

Answer:

$$x[n] = \frac{1}{N}\sum_{k=0}^{N-1} X[k]\,.\,e^{j2\pi kn/N}$$

Substituting the values:

$x\,[0] = \frac{1}{4}\,(3\,.e^{j0} + (1+j)\,.e^{j0} - 2\,.e^{j0} + (1-j)\,.e^{j0})$

$= \frac{1}{4}(3+1+j-2+1-\mathrm{j}) = \frac{1}{4}(3+1-2+1) = \frac{1}{4}(3) = \frac{3}{4}$

$x\,[1] = \frac{1}{4}\,(3.\,e^{j2\pi/4} + (1+j).\,e^{j2\pi/4} - 2.\,e^{j2\pi/4} + (1-j)\,.e^{j2\pi/4})$

$= \frac{1}{4}(3-3j-2-2j+1-1j) = \frac{1}{4}(3-2+1-3j-2j-1j)$

$= \frac{1}{4}\,(2-6j)$

$x\,[2] = \frac{1}{4}\,(3.\,e^{j4\pi/4} + (1+j).\,e^{j4\pi/4} - 2.\,e^{j4\pi/4} + (1-j)\,.e^{j4\pi/4})$

$= \frac{1}{4}(3-1-j-2-2j+1+1j) = \frac{1}{4}(3-2-1-2j+1j)$

$= \frac{1}{4}(0-j) = -\frac{j}{4}$

$x\,[3] = \frac{1}{4}\,(3.\,e^{j6\pi/4} + (1+j).\,e^{j6\pi/4} - 2.\,e^{j6\pi/4} + (1-j)\,.e^{j6\pi/4})$

$= \frac{1}{4}(3+3j-2+2j+1+j-1+j) = \frac{1}{4}(3-2+1+3j+2j+1j-1+1j)$

$= \frac{1}{4}(1+6j)$

So, $x[n] = \{\frac{3}{4}, \frac{1}{2} - \frac{3j}{4}, -\frac{j}{4}, \frac{1}{4} + \frac{3j}{2}\}$

Example 4.9. If a continuous-time signal (t) has a Fourier transform (f) with a bandwidth of 20 kHz, what is the Nyquist frequency for sampling this signal?

Answer:

The Nyquist frequency (f_N) is half the sampling frequency (f_s). According to the Nyquist theorem, to avoid aliasing, the sampling frequency must be at least twice the maximum frequency component in the signal.

Since the signal has a Fourier transform with a bandwidth of 20 kHz, the maximum frequency component in the signal is f_{max} = 20 kHz. Therefore, according to the Nyquist theorem, the sampling frequency (f_s) must be at least 2×20 kHz.

So, the Nyquist frequency (f_N) is given by:

$f_N = f_s/2$

Substituting f_{max} = 20 kHz:

f_N= (2×20 kHz)/2 = 20 kHz

Therefore, the Nyquist frequency for sampling this signal is f_N= 20 kHz.

Example 4.10. If $(\omega)=4\delta(\omega-\omega_0)+3\delta(\omega-\omega_0)$, where $\delta(\omega)$is the Dirac delta function, determine the time-domain signal $x(t)$.

Answer:

To find the time-domain signal (t) corresponding to the given frequency-domain signal (ω), we can use the inverse Fourier transform. The inverse Fourier transform of X(ω) gives us $x(t)$.

$X(\omega)=4\delta(\omega-\omega_0)+3\delta(\omega-\omega_0)$

We know that the inverse Fourier transform of a Dirac delta function $\delta(\omega-\omega_0)$ is a complex exponential $e^{j\omega 0t}$, and the inverse Fourier transform of $\delta(\omega+\omega 0)$ is $e^{-j\omega 0t}$.

So, using the linearity property of the inverse Fourier transform, the time-domain signal $x(t)$ is:

$x(t) = F^{-1}\{4\delta(\omega-\omega_0)\}+F^{-1}\{3\delta(\omega-\omega_0)\}$

$x(t) = 4e^{j\omega 0t}+3e^{j\omega 0t}$

$x(t) = 4e^{j\omega 0t}+3e^{j\omega 0t}$

$x(t) = (4\ e^{j\omega 0t}+3\ e^{j\omega 0t}) = (4+3)\ e^{j\omega 0t}$

$x(t) = 7e^{j\omega 0t}$

So, the time-domain signal $(t)=7e^{j\omega 0t}$

Example 4.11. A signal (t) has a spectrum with non-zero values only at f=±50 Hz. What is the bandwidth of this signal?

Answer:

The bandwidth of the signal is the range of frequencies over which it has non-zero values. In this case, the bandwidth is 100 Hz.

Example 4.12. If a continuous-time signal (t) has a Fourier transform (f) with significant energy only in the frequency range from 100 Hz to 500 Hz, what is the minimum sampling frequency required to accurately represent this signal in the discrete-time domain?

Answer:

To accurately represent a continuous-time signal in the discrete-time domain without aliasing, we need to sample it at a frequency high enough to capture all the significant frequency components. According to the Nyquist theorem, the sampling frequency (f_s) must be at least twice the maximum frequency component in the signal.

Given that the significant energy of the signal lies in the frequency range from 100 Hz to 500 Hz, the maximum frequency component (f_{max}) is 500 Hz.

So, according to the Nyquist theorem, the minimum sampling frequency (f_{min}) required to accurately represent the signal is:

$f_{min} = 2 \times f_{max}$

Substituting the given value $f_{max} = 500$ Hz:

$f_{min} = 2 \times 500 \text{ Hz} = 1000 \text{ Hz}$

Therefore, the minimum sampling frequency required to accurately represent this signal in the discrete-time domain is $f_{min} = 1000$ Hz

Example 4.13. If a signal (t) has a Fourier transform (f) and its spectrum is symmetric around $f = 0$, what can you conclude about the signal $x(t)$?

Answer:

A symmetric spectrum around $f = 0$ implies that the signal (t) is real-valued.

Example 4.14. Determine the frequency resolution of a Discrete Fourier Transform (DFT) analysis of a discrete-time signal with a sampling frequency of 1000 Hz and a DFT size of 1024 points.

Answer:

The frequency resolution of a DFT analysis is determined by the spacing between adjacent frequency bins in the frequency domain. It is calculated as the sampling frequency divided by the DFT size.

Given:

Sampling frequency $(f) = 1000$ Hz

DFT size = 1024 points

The frequency resolution (Δf) is given by:

$\Delta f = f_s / N$

where:

f_s is the sampling frequency

N is the DFT size

Substituting the given values:

$\Delta f = 1000 \text{ Hz} / 1024$

$\Delta f \approx 0.976$ Hz

So, the frequency resolution of the DFT analysis of the given discrete-time signal is approximately 0.976 Hz.

Example 4.15. If a signal (t) is band-limited to frequencies between 0 Hz and 2000 Hz, what is the minimum sampling frequency required to reconstruct the signal without loss of information?

Answer:

The minimum sampling frequency required to reconstruct a band-limited signal without loss of information is determined by the Nyquist theorem. According to the Nyquist theorem, the sampling frequency (f) must be at least twice the maximum frequency component in the signal.

Given that the signal (t)) is band-limited to frequencies between 0 Hz and 2000 Hz, the maximum frequency component $(f\text{max})$ is 2000 Hz.

So, according to the Nyquist theorem, the minimum sampling frequency $(f\text{min})$ required to reconstruct the signal without loss of information is:

$f\text{min}=2\times f\text{max}$

Substituting the given value $f\text{max} = 2000$ Hz:

$f\text{min} = 2\times2000 \text{ Hz} = 4000 \text{ Hz}$

Therefore, the minimum sampling frequency required to reconstruct the signal (t) without loss of information is $f\text{min} = 4000$ Hz.

Example 4.16. The spectrum of a signal x(t) has a peak at f=200 Hz with a magnitude of 20. What is the amplitude of the corresponding cosine component in the time domain?

Answer:

The amplitude of the cosine component in the time domain is half of the peak magnitude in the frequency domain, so it is 10.

Example 4.17. If a signal has a period of $T = 0.02$ seconds, what is its fundamental frequency?

Answer :

The fundamental frequency f_0 is the reciprocal of the period, so $f_0 = 1/T = 1/0.02 = 50$ Hz.

Example 4.18. If a signal x(t) has a Fourier transform $X(f)$ with a peak magnitude of 10 at f=100 Hz, what is the peak magnitude in the time domain?

Answer:

The peak magnitude in the time domain is equal to the value of (f) at f=100 Hz, which is 10.

Example 4.19. Given a continuous-time signal x(t)=cos($2\pi ft$), where f=100 Hz, find its Fourier Transform $X(f)$.

Answer:
The Fourier Transformation $X(f)$ of a continuous-time signal $x(t)$ is defined as

$$X(f) = \int_{-\infty}^{\infty} x(t).e^{-j2\pi ft}dt$$

Given the continuous-time signal $x(t) = \cos(2\pi ft)$, where $f = 100$ Hz, let's find its Fourier Transform $X(f)$:

$x(t) = \cos(2\pi ft)$

Substitute $f = 100$Hz:

$x(t) = \cos(2\pi.100.t)$

Using Euler's formula, $\cos(\theta) = \frac{e^{j\theta} + e^{-j\theta}}{2}$, we can rewrite x (t) as:

$$x(t) = \frac{e^{j2\pi.100.t} + e^{-j2\pi.100.t}}{2}$$

Now, let's find the Fourier Transform $X(f)$:

$$X(f) = \int_{-\infty}^{\infty} \frac{e^{j2\pi.100.t} + e^{-j2\pi.100.t}}{2}.e^{-j2\pi ft}dt$$

$$X(f) = \frac{1}{2}\int_{-\infty}^{\infty} e^{j2\pi.100.t}.e^{-j2\pi.100.t}dt + \frac{1}{2}\int_{-\infty}^{\infty} e^{j2\pi.100.t}.e^{-j2\pi.100.t}dt$$

$$X(f) = \frac{1}{2}\int_{-\infty}^{\infty} e^{j2\pi.100.t-j2\pi.100.t}dt + \frac{1}{2}\int_{-\infty}^{\infty} e^{j2\pi.100.t-j2\pi.100.t}dt$$

$$X(f) = \frac{1}{2}\left(\int_{-\infty}^{\infty} e^{j2\pi.100.t-j2\pi.100.t}dt + \int_{-\infty}^{\infty} e^{j2\pi.100.t-j2\pi.100.t}dt\right)$$

$$X(f) = \frac{1}{2}(\delta(f-100) + \delta(f+100))$$

So, the Fourier Transform X (f) of the given continuous time signal x (t) = cos (2πf t), where f = 100 Hz, is:

$$X(f) = \frac{1}{2}(\delta(f-100) + \delta(f+100))$$

Example 4.20. Given a signal $(t) = 2\cos(2\pi\cdot 10t) + 3\sin(2\pi\cdot 20t)$, what are the frequencies present in $x(t)$?

Answer:
The frequencies present in (t) are 10 Hz and 20 Hz.

REFERENCES AND SUGGESTED READINGS

1. Santamarina, J. C. (2005). Discrete signals and inverse problems : an introduction for engineers and scientists. Wiley, Hoboken, NJ..
2. Oktay, Z. N. (2021). Digital filter and analog LDO design for ADPLL in 180 nm CMOS technology (Master's thesis, İstanbul MedipolÜniversitesi Fen BilimleriEnstitüsü).
3. Wang, H. (2009). FFT Basics and Case Study using Multi-Instrument. Rep. No. Rev,1.
4. Handouts and Mathcad scripts from Dr. Brina Montoya, NCSU.
5. Internet sources

KNOW MORE

For more information related to this topic scan the QR code.

OR

Type this link in your browser

https://www.youtube.com/watch?v=jsgnVKXvpXw

CO-PO Attainment Table

Course outcomes (COs) for this course can be mapped with the programme outcomes (POs) after the completion of the course and a correlation can be made for the attainment of POs to analyze the gap. After proper analysis of the gap in the attainment of POs necessary measures can be taken to overcome the gaps.

Table for CO and PO attainment

Course Outcomes	Attainment of Programme Outcomes (1- Weak Correlation; 2- Medium correlation; 3- Strong Correlation)											
	PO-1	PO-2	PO-3	PO-4	PO-5	PO-6	PO-7	PO-8	PO-9	PO-10	PO-11	PO-12
CO-1												
CO-2												
CO-3												
CO-4												
CO-5												
CO-6												
CO-7												
CO-8												

The data filled in the above table can be used for gap analysis.

INDEX